THE INSECTS AND ARACHNIDS OF CANADA

PART 1

Collecting, Preparing, and Preserving Insects, Mites, and Spiders

Compiled by

J. E. H. Martin

Biosystematics Research Institute
Ottawa, Ontario

Research Branch
Canada Department of Agriculture

Publication 1643   1977

©Minister of Supply and Services Canada 1978

Available by mail from

Printing and Publishing
Supply and Services Canada
Hull, Québec, Canada K1A 0S9

or through your bookseller.

Catalogue No. A42-42/1977-1 Canada: $3.50
ISBN 0-660-01650-8 Other countries: $4.20

Price subject to change without notice.

Printed by Kromar Printing Ltd.
02KT.01A05-7-38911

The Insects and Arachnids of Canada.

Part 2. The Bark Beetles of Canada
and Alaska (Coleoptera: Scolytidae),
by Donald E. Bright, Jr.,
Biosystematics Research Institute,
Ottawa, was published in 1976.

Part 3. The Aradidae of Canada
(Hemiptera: Aradidae),
by Ryuichi Matsuda,
Biosystematics Research Institute,
Ottawa, was published in 1977.

Part 4. The Anthocoridae of Canada
and Alaska (Heteroptera: Anthocoridae),
by Leonard A. Kelton,
Biosystematics Research Institute,
Ottawa, is in press.

Contents

Foreword	9
Introduction	11
Equipment and methods for collecting	11
General-purpose nets	11
Nets for flying insects	14
Lights and light traps	15
Baits and bait traps	20
Other traps for flying insects	23
Malaise traps	23
Window traps	28
Tent-window traps	28
Visual-attraction traps	30
Pan traps	30
Baits	31
Sweeping	33
Beating	34
Emergence cages and separators	40
Searching	42
Barking	42
Mammal nests	43
Aspirator	43
Collecting leafmining insects	46
Collecting and extracting arthropods from debris	47
Berlese funnel	47
Sifter and photoeclector	54
Collecting and extracting burrowing and boring arthropods	56
Collecting aquatic insects and mites	57
Aquatic traps	63
Lentic environment traps	63
Surface emergence traps	64
Tow traps	65
Benthonic samplers	65
Lotic environments	66
Collecting ectoparasites of vertebrates	67
Collecting mites associated with invertebrates	69
Rearing	70
Killing agents and killing bottles	73
Equipment and methods for preserving and mounting	79
Relaxing	79
Cleaning	81
Temporary storage in papers	82
Temporary storage by refrigeration	84
Temporary storage by layering	85
Direct pinning	85
Double mounting	90
Spreading	90
Adhesives and pointing	94
Preservation in liquids	98
Dry preservation of soft-bodied insects	101
Microscopical preparations	102
Insects	102

Mites	107
Data labels	109
Storage and care of collections	113
Storage and care of pinned insects	113
Storage of a collection preserved in liquid	116
Storage of microscope slides	116
Packing insects and arachnids for shipment	118
Studying the collection	122
Applying the methods	124
Thysanura	124
Diplura	125
Collembola and Protura	125
Orthoptera and Dermaptera	125
Plecoptera	128
Isoptera	128
Corrodentia	128
Mallophaga	128
Anoplura	129
Ephemeroptera	130
Odonata	131
Thysanoptera	132
Hemiptera (including Heteroptera and Homoptera)	133
Preserving and mounting Hemiptera	134
Preserving and mounting Aphidoidea, Coccoidea, and Aleyrodoidea	135
Megaloptera	136
Neuroptera	137
Mecoptera	137
Trichoptera	137
Lepidoptera	138
Coleoptera	142
Strepsiptera	145
Hymenoptera	146
Symphyta	146
Ichneumonoidea	147
Microhymenoptera	148
Apoidea, Sphecoidea, Vespoidea, Scolioidea	151
Formicidae	151
Diptera	152
Collecting adults	152
Assembly sites	153
Artificial light	153
Sunlight	153
Traps	153
Chemicals	154
Unbaited traps	154
Pan traps	155
Preserving and mounting	155
Pinning	155
Freeze-drying	156
Double mounts	156
Pinning from alcohol	156
Dissected genitalia	157
Microscopic mounts	157
Collecting and rearing immature stages	157

       Aquatic larvae 158
       Leafminers 158
       Soil-dwelling larvae 158
       Larvae in decaying vegetation 159
       Carrion and dung feeders 159
       Parasites 159
       Aphid predators 159
       Larval and pupal exuviae 159
   Siphonaptera 161
   Acari (mites) 162
   Ixodides 167
   Araneae 167
Formulas 169
References 172
Index 175

Foreword

This publication replaces *Collecting, Preparing and Preserving Insects,* Canada Department of Agriculture Publication 932, compiled by Bryan P. Beirne and published in 1955. This text was prepared by members of the zoology staff of the Biosystematics Research Institute, except for a contribution on the order Orthoptera by Dr. V. R. Vickery of Macdonald College, McGill University, Montreal, Que.

This guide to collecting and preserving has a dual objective: to acquaint amateur entomologists or potential amateurs with the basic methods of obtaining insect and arthropod material and preparing a collection; and to serve as a guide to both amateur and professional entomologists on how to prepare material before submitting it to the National Identification Service of the Biosystematics Research Institute. Address your inquiries regarding the identification of insects to:

Officer-in-Charge
National Identification Service
Biosystematics Research Institute
Agriculture Canada
Ottawa, Ontario K1A 0C6

This publication is Part 1 of the Canadian faunal series, *The Insects and Arachnids of Canada.* The series will treat various groups of arthropods occurring in Canada. Difficulty with the preparation of suitable illustrations delayed the publication of this volume, and therefore it has been preceded by Part 2 in the series, *The Bark Beetles of Canada and Alaska,* which was published in 1976. Several additional manuscripts have been completed and will soon be published as further parts of the Canadian faunal series.

Contributions by the following people to the preparation of this book are acknowledged. John E. H. Martin assumed the task of compiling the various submissions and of preparing several of the chapters. Dr. E. E. Lindquist and G. E. Shewell made extensive contributions to the text. Susan Rigby and Bernard Baker prepared the illustrations. Drs. G. P. Holland, E. C. Becker, and W. R. Richards reviewed the manuscript in its early stages, and the Editorial Committee of the Canadian faunal series, Drs. D. E. Bright, C. D. Dondale, and D. R. Oliver, reviewed the final typescript. D. M. Archibald of the Research Program Service made a material contribution to this volume by her constructive editing of the manuscript.

D. F. Hardwick
Director
Biosystematics Research Institute

Introduction

In entomology the main reason for using various collecting methods and equipment is to obtain specimens of all kinds of insects, mites, and spiders in the easiest and most effective way. The chief aim of using various methods and equipment for preparing and preserving insects is to preserve the specimens in a natural, undamaged, and undistorted condition. The methods and equipment described in this publication have been found satisfactory. However, because none accomplishes its objective completely, each one may be improved or adapted to new conditions. Moreover, new methods and equipment are continually being developed. To assist in improving and developing new techniques, the principles, advantages, and disadvantages of the best-known methods and equipment are described here.

Equipment and methods for collecting

Various methods of collecting may be used to collect insects from each of the 10 environmental situations in which insects occur. Insects are found in the air; in fresh or brackish water; on the foliage or stems of trees and shrubs; on low-growing plants; on the ground or near the roots of low-growing plants; internally in plants, plant products, seeds, and fruit; among debris; in the nests or habitations of animals and man; in the soil; and on or in other insects or animals. Because certain collecting methods and situations or environments at times may produce insects in large numbers, they are preferred to the methods, situations, and times that are less productive quantitatively but that may produce species that otherwise are rarely captured. The latter methods must be used, however, in order to complete a survey of the insect fauna of a region, locality, habitat, or plant, or a collection of a particular insect group.

Collecting may be merely a matter of picking up the insects with your fingers after you have watched their habits and discovered their habitats. However, obtaining specimens quickly and in large numbers often requires extensive field observations and special collecting equipment. Some kinds of equipment operate on the principle of extracting the insects manually or mechanically from the situations in which they occur; others concentrate or trap the insects by taking advantage of their normal movements or of their reactions to light, gravity, heat, moisture, and odors.

General-purpose nets

Nets are useful for catching flying and aquatic insects and for sweeping up insects from vegetation. If you intend to collect by only one of these methods, then use a net designed for that specific method. These specially designed nets are described under their appropriate headings: "Nets for Flying Insects" on p. 14, "Collecting Aquatic Insects and Mites" on p. 57, and

"Sweeping" on p. 33. The differences in design consist of modifications of some or all of the three main parts of the net: the ring, the bag, and the handle. General-purpose nets can be used for various kinds of collecting by merely changing the bag.

Use a net with a ring that detaches from the handle so that the bag can be replaced easily when it wears out, gets wet, torn, dirty, or has to be changed for one of another material. A folding ring (Fig. 1) is easy to carry when you are traveling.

A simple net (Fig. 2) has a 38 cm (15 in.) ring of about 3 mm (1/8 in.) iron or steel wire. The ends of the wire are straight and fit into grooves in the handle; their tips are bent inward to fit into holes in the handle. The ends may be held in place by binding them firmly with insulating or adhesive tape or by using a sliding metal sleeve. The use of a sliding metal sleeve makes the ring easier to remove from the handle.

Perhaps the best net for general purposes is the kind that is used as a fisherman's landing net (Fig. 3), though you may find it heavy to carry. The ring is made of spring steel bands and is collapsible. It opens to about 33–38 cm (13–15 in.) in diam. The ring screws into a ferrule at the top of the handle; it can easily be taken apart to change the bag by unscrewing a bolt in the ring. The rings of many landing nets are too flimsy for use as insect nets, but those made of heavier steel are satisfactory if you strengthen the joints with solder or extra rivets.

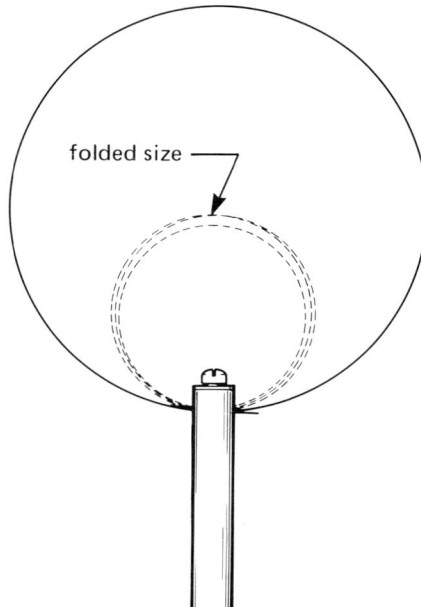

Fig. 1. An insect net that can be folded and carried in a pocket by twisting the top of the net ring and the handle in opposite directions.

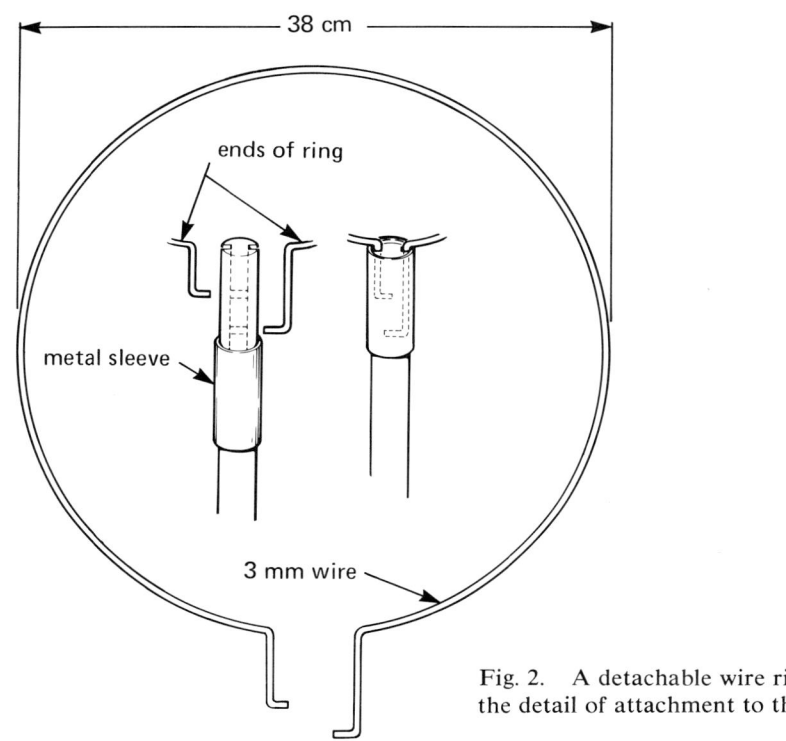

Fig. 2. A detachable wire ring showing the detail of attachment to the handle.

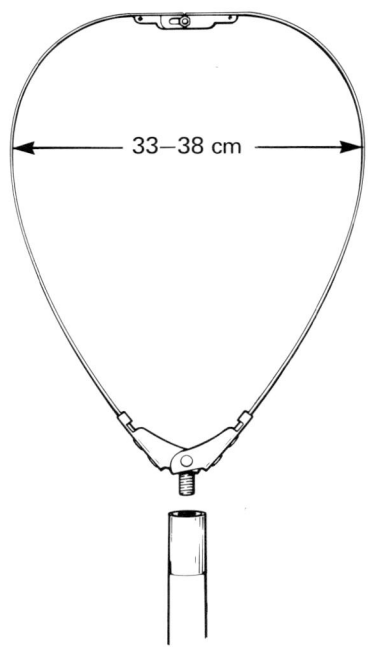

Fig. 3. A folding flat metal ring.

The handle of the net should be strong and light. A bamboo handle or one of 19 mm (3/4 in.) hardwood dowel is suitable. A joined wooden handle or one made from an aluminum ski pole is adaptable for many purposes, but it may be too heavy. For most purposes, a handle 0.6–0.9 m (2 or 3 ft) long is most convenient.

The kind of material used for the bag should depend on the method and purpose of collecting. A net bag should have its edge, where it is attached to the ring, made of strong cloth such as light canvas, heavy muslin, or linen. Fold this strong cloth over to form a hem about 7 cm (3 in.) deep so that you can pass the ring through it. The hem increases the strength of the bag around the rim where the greatest wear takes place. Also, a bag made in this way is easier to change when you are in the field.

Nets for flying insects

Nets for catching flying insects must be as light as possible, with the least possible air resistance, and yet be reasonably strong and durable. A 38 cm (15 in.) ring is satisfactory, because a smaller net is less air resistant than a larger one. The rings described previously are suitable for use with aerial nets.

The length of the handle may vary, depending on your preference and also on the kind of insects being collected. A long handle is a great help in catching high-flying butterflies and day-flying or dusk-flying moths; a short handle is most convenient for small, active insects.

Make the bag about twice or two and one-half times as deep as the diameter of the ring, or slightly shorter than your arm. You can make the bag in two sections (Fig. 4), each somewhat tapered toward the bottom, which is broadly rounded. If you make the bag of four such sections, it opens better when you are using it.

Choose a fabric for the bag that is a light and fairly transparent net material with little air resistance. The meshes may be as large as possible, but small enough to hold the insects you want to capture. Woven fabric is usually unsuitable for net bags, because it is too stiff and often damages the wings, especially of Lepidoptera. However, mosquito netting is suitable if you wash it thoroughly to remove the stiffness. The best material is one in which the threads are not woven but are twisted together when they meet. The standard material is Brussels netting, brusselette, or bobbinet; these fabrics have hexagonal meshes that do not lose their size or shape under normal stress. Choose a cotton material and be sure to wash it before you use it. Do not use rayon bobbinet because it is too weak. Organdy is an excellent material to use for several reasons: it is cheap, easily obtainable, and durable, does not bunch up or scratch the wings of Lepidoptera, dries quickly, and may be used with some success even when it is wet. The disadvantages of using organdy are its higher resistance to air and its lower transparency than some other materials. Other fabrics suitable for net bags are nylon, marquisette, tulle, and good-quality cheesecloth.

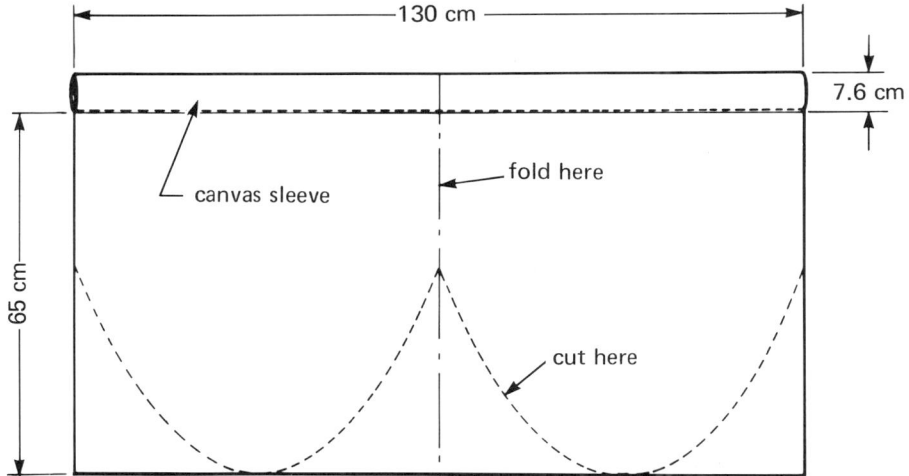

Fig. 4. A pattern for making a net insect bag.

Remove small insects (except Lepidoptera) from the net with a killing bottle or aspirator. When you catch a large insect or a specimen of Lepidoptera in your net, twist the handle quickly and lap the bag over the ring, enclosing the specimen in the bottom of the bag. To remove the insect from the net, grasp the bag and enclose the specimen in a small fold of the cloth. With your other hand, insert an open killing bottle into the net and push it upward until it encloses the specimen. Put your hand over the top of the bottle and withdraw the bottle from the net and cork it. If the specimen is very active or capable of stinging, insert the fold of the net containing the insect into the killing bottle until the insect is stupefied.

Lights and light traps

The most productive way of collecting nocturnal flying insects is to take advantage of their strong attraction to artificial light. You can capture the insects with a net or a killing bottle as they fly around the light or rest nearby, or you can use traps that capture and hold them as they approach the light.

Insects come to light usually in largest numbers on still, dark, cloudy nights when both temperature and humidity are high, and in smallest numbers on cold, windy, moonlit nights, though almost any night may be good. The use of a mercury-vapor lamp, black light, or other lamp high in ultraviolet results in the capture of very large numbers of specimens of some species that are rarely or never attracted by ordinary white light, and of females, which are often scarce at ordinary light. There is some evidence that the higher the surface brilliance of the light source the larger the number of insects attracted.

Light does not attract insects from long distances; it attracts only those that are already in or that enter the illuminated area.

The usual method of collecting in the field is to put a light (a generator- or battery-powered 10 or 15 W black-light fluorescent tube) on or over a white sheet laid on the ground or to suspend it in front of a vertical sheet. The latter produces better results; stretch the sheet between two posts held up by guy ropes, suspended from a branch of a tree or from a rope or wire, or attached to a wall or fence. In open country where sudden gusts of wind may sweep the insects off the sheet or even blow the sheet away, arrange the sheet so that it is quickly removable.

The use of light traps is one of the most efficient methods of collecting specimens and species in large numbers. However, it has some disadvantages. On a good night, large quantities of insects accumulate in a trap and may damage one another; the wings of moths become broken and rubbed, and other insects become covered with scales and hairs from the moths. If you are using a trap to obtain specimens for a collection, clear out the container or killing bottle at frequent intervals, before a large number of insects has accumulated in it. Or, use a powerful killing agent that will stupefy the specimens almost immediately. The latter is not usually recommended because of the poison danger to humans. Moreover, it is often difficult to keep the fumes at a high concentration, because they dissipate through the entrance of the trap. You can place a dish of alcohol below a light to catch insects that are not harmed by immersion in alcohol; this method is unsuitable for Lepidoptera. If you are collecting small insects other than Lepidoptera, place a wire screen with mesh of a suitable size around or below the light to filter out the larger specimens and to leave the smaller ones undamaged.

To make a simple funnel trap (Fig. 5), suspend a light over a metal or stiff-paper funnel. The insects fall down the funnel into a killing bottle below. If you want to keep the insects alive, replace the killing bottle with a large cylinder of cheesecloth suspended from the rim of the funnel. The 10 or 15 W black-light trap (Fig. 6) is based on the same principle, but has a larger metal container, a metal baffle, and a funnel. The large container in place of a killing bottle provides plenty of room for the captures to fly around, and thus reduces the likelihood of damage; and it can accommodate the larger numbers captured when you use an ultraviolet lamp.

A funnel trap is efficient for capturing heavy-bodied or fast-flying insects, such as noctuid moths. The insects strike the light or baffle plates and fall into the funnel. It is less efficient for capturing light-bodied or weak-flying insects, because they enter the funnel only accidentally in their random flights or fall into it if partly overcome by the heat of the lamp or the fumes of the killing agent. However, the kinds of insects captured vary, depending on the type of light used.

A fluorescent black light on or over a white sheet is more efficient for capturing beetles and certain other insects. Often these insects fall short of the trap and stay on the ground.

Fig. 5. A simple funnel light trap.

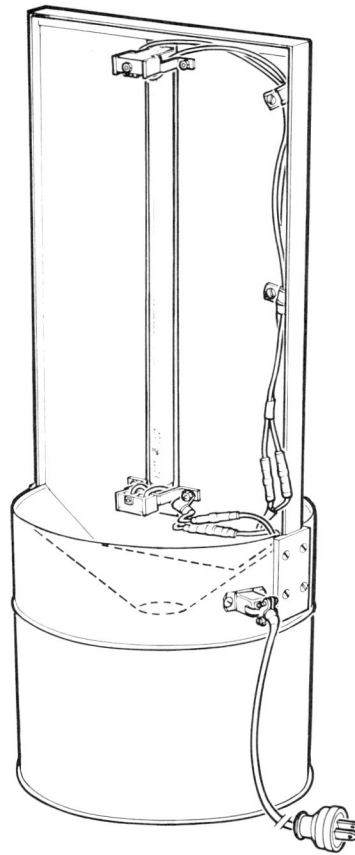

Fig. 6. A 10 or 15 W black-light trap.

A specialized funnel trap (Fig. 7) has been used successfully for collecting noctuid moths for several years. The outer shell of the trap is a 51 cm (20 in.) high galvanized steel garbage can to which other components have been fitted.

The light source (*a*) is a 125 W 200–220 V Osram mercury-vapor globe. It is enclosed by four baffles (*b*), which extend slightly above the level of the top of the bulb. The sharply sloped funnel (*c*) below the light leads into an inner metal chamber (*d*). The inner chamber, which confines as much as possible of the gas generated, may be lifted out of the shell of the trap when the funnel is removed. Inside the inner chamber and just below the lower end of the funnel is a small container with a screen lid, the rain drain (*e*), which has a tube leading from it down through the bottom of the outside of the trap; the container catches and drains off any water that enters through the funnel. The screened lid of the rain drain must be close enough to the lower end of the funnel and wide enough in diameter to prevent rain from dripping on the floor of the reception chamber. A circular 13 mm (1/2 in.) thick disk of sponge rubber fastened to the upper surface of the screened lid of the rain drain cushions the fall of larger noctuids as they drop down into the trap at a sharp angle after entering at high speed. Noctuids may enter the trap at such an angle that they hit the cover of the rain drain so hard that a cloud of ascending scales can be seen above the funnel of the trap. Below the basin of the rain drain in the reception chamber is a removable metal tray with a 3 mm (1/8 in.) mesh hardware cloth bottom; the floor of the tray is covered with a thin pad of cheesecloth. A removable metal lattice that divides the tray into several compartments rests on the cheesecloth. Below the metal tray, on the floor of the reception chamber, is a 13 mm (1/2 in.) thick pad of cheesecloth saturated with the killing agent tetrachloroethane. A 100 W heating element, which vaporizes the tetrachloroethane and warms the reception chamber to keep the chemical vaporized, is located in a separate chamber at the bottom of the trap. The heating element is separated from the pad containing the tetrachloroethane only by the thickness of the metal forming the bottom of the reception chamber.

To disassemble the trap in order to inspect the catch, first remove the funnel, then lift the reception chamber from the outer shell and remove its lid.

Remove the rain drain so that the tray containing the night's catch can be lifted out of the killing chamber. In a well-ventilated room, sort and pin the specimens directly from the tray. When you reassemble the trap before reusing it, dampen the cheesecloth pad on the bottom of the reception chamber with 40–50 cm^3 of tetrachloroethane.

On nights when you expect particularly heavy flights, increase the concentration of tetrachloroethane vapor in the reception chamber by inserting a wad of cheesecloth in the basin of the rain drain and saturating it with the killing agent. A common problem with a trap that has a large reception chamber is that in dry locations the specimens dry out before they can be sorted and pinned. You may find it helpful to keep the humidity high in the reception chamber by adding water (about the same amount of water as tetrachloroethane) to the pad at the bottom of the chamber and to the wad

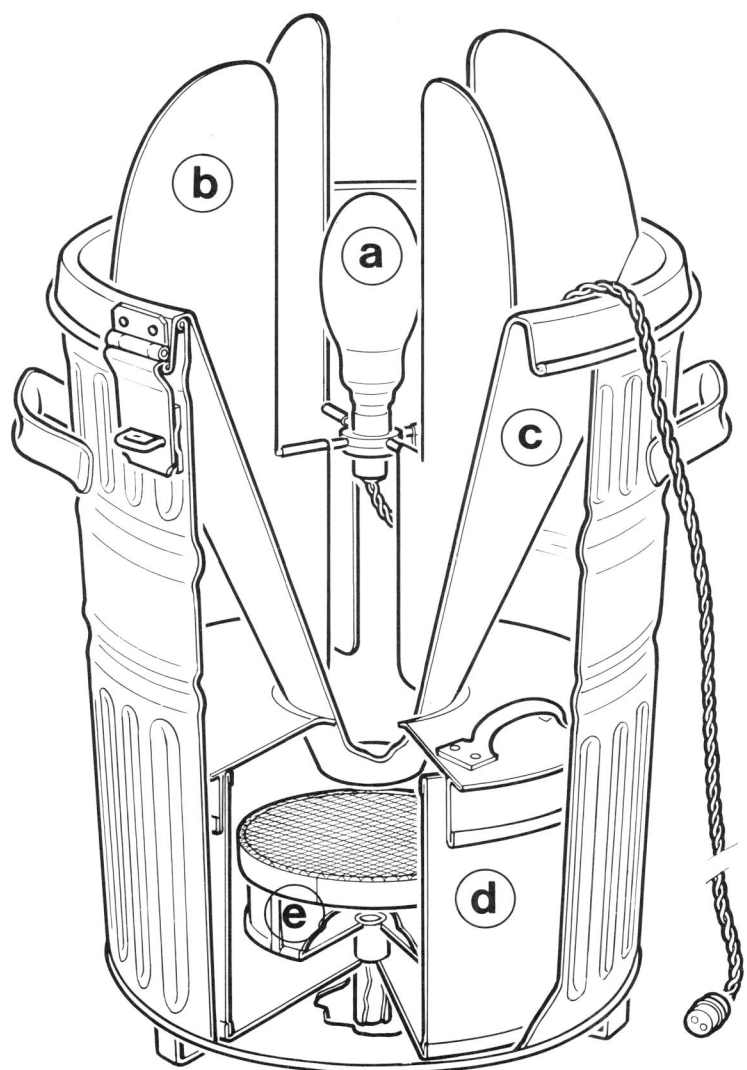

Fig. 7. A mercury-vapor trap: (*a*) 125 W 200-220 V Osram mercury-vapor globe, (*b*) four metal baffles, (*c*) metal funnel, (*d*) metal inner killing chamber, (*e*) rain drain with tube leading to the outside of the trap.

of cheesecloth in the basin of the rain drain. In areas where electrical power is not available for operating the light on the trap, use a Coleman lamp.

A box trap (Fig. 8) may be more effective than a funnel trap for some insects. Most of the insects attracted, including the light-bodied and feebly flying ones, enter and are captured. The box trap is more effective because the insects are trapped before they come close to the light. The efficiency of a box trap increases with its size (that is, with the distance of the entrance from the light). However, it is much less efficient than a funnel trap. A major disadvantage is that a box trap attracts insects from only one direction.

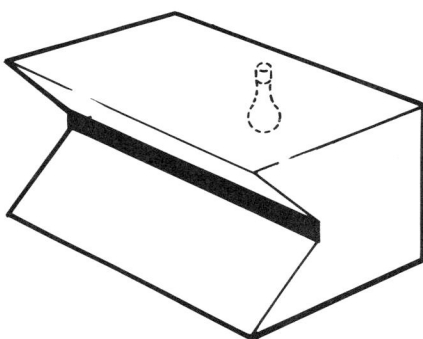

Fig. 8. A simple box trap.

Tetrachloroethane, carbon tetrachloride, or other liquid killing agents may be used in a funnel trap that has a large container. An open flat tin filled with plaster of Paris or some other absorbent material, saturated with the liquid and placed in the container, produces an effective concentration of vapor. A jar half-filled with alcohol may be used instead of a gaseous killing agent for insects that are not harmed by immersion in alcohol. In addition to killing the specimens, the alcohol washes off lepidopterous scales and hairs.

Light, bait, and other traps, depending on their design, location, and the way in which they are used, can provide a great deal of information on the habits of insects (for example, the times of day or night when they are active, their relative abundance, and their flying heights).

Baits and bait traps

Many insects are attracted to and congregate at suitable food, either for themselves or for their larvae or nymphs. Take advantage of these natural baits by concentrating them or putting them in convenient situations or in traps, or by simulating them with artificial baits. Different natural baits are attractive to different kinds of insects. Some natural baits used by collectors are flowers, the aphid excretion known as honeydew, the fermenting sap that exudes from wounds in trees, rotting or overripe fruit, carrion, animal excreta, and rotting fungi.

A jar, tin can, or larger container, with or without a cone, may be used as a pitfall (Fig. 9) to catch beetles and other insects that crawl on the surface of the ground. Bury the container to its rim in the soil, and the insects that fall in are unable to get out. Put a piece of wood or a flat stone over the mouth of the pitfall to keep out rainwater, but leave enough space for the insects to enter. Check pitfall traps often to prevent insects from damaging each other by crawling over one another. When checking is not possible, it is better to suspend the bait in the can or jar in a cheesecloth bag from a stick or 13 mm (1/2 in.) mesh hardware cloth screen. To kill and preserve the insects, place a solution of equal amounts of water and ethylene glycol below the bait. This solution drowns the insects and prevents their decomposition in temperate areas for at least a week. When you use the large 13 mm (1/2 in.) mesh screen, weight it on the edges with rocks or logs to prevent mammals from stealing the bait. Large numbers of traps can be packed into a small space and taken on trips if they are waxed paperboard or plastic cups or refrigerator boxes that fit into one another.

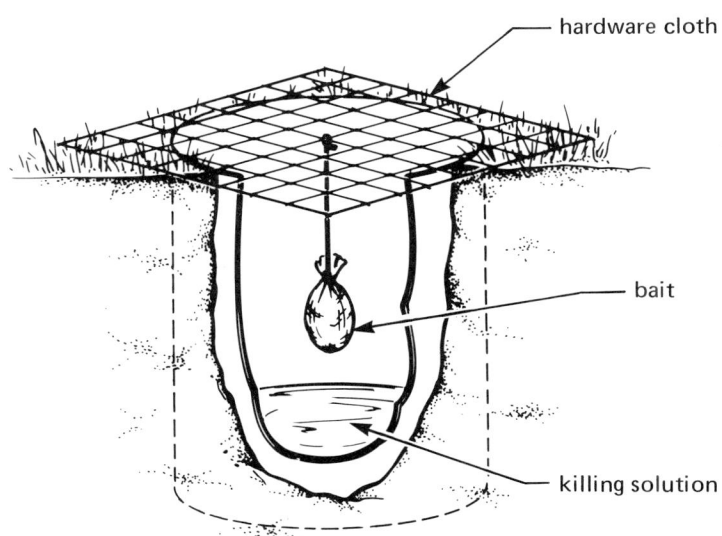

Fig. 9. A pitfall trap with suspended bait.

The types of bait traps for flying insects are shown in Figs. 10 and 11. In one trap (Fig. 10) the bait is placed in the pan and the insects enter through the opening below the cylinder. When they have fed, they fly upward and pass through the opening at the apex of the cone into the container. The other trap (Fig. 11) is designed to operate on the same principle as a funnel light trap, except that the lamp is replaced by a wire screen bait container. A simplified modification of this design is an inverted cone with an opening at its apex placed in the mouth of a baited jar.

Artificial baits have not been investigated thoroughly, though they might be effective and profitable. One bait that is used extensively is sugar; it is used to attract nocturnal Lepidoptera (see "Lepidoptera," p. 138).

Fig. 10. A funnel bait trap.

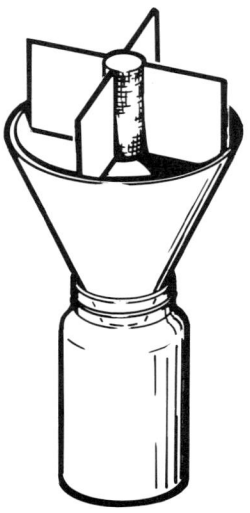

Fig. 11. A cone bait trap.

Other traps for flying insects

Some kinds of traps are effective without attractants for the insects. There are two types: traps that move to trap the insects and traps that the insects enter accidentally. Some of these traps have certain advantages over other methods of collecting flying insects: they operate continuously, and therefore they capture some insects that fly when collectors are not usually present; and they capture some flying insects that are too small to be seen or netted easily and that fly so high above the vegetation that they tend to escape capture by the more conventional collecting methods.

Malaise traps The Malaise trap (Figs. 12–15) is based on the principle that most insects fly or climb upward when confronted with a barrier and also move from a dark to a lighter area. Weak fliers or insects that drop when they meet an obstruction are seldom taken in a Malaise trap. However, Malaise traps are especially effective in collecting Diptera, Hymenoptera, and Lepidoptera, but less useful in collecting Coleoptera, Heteroptera, and Homoptera.

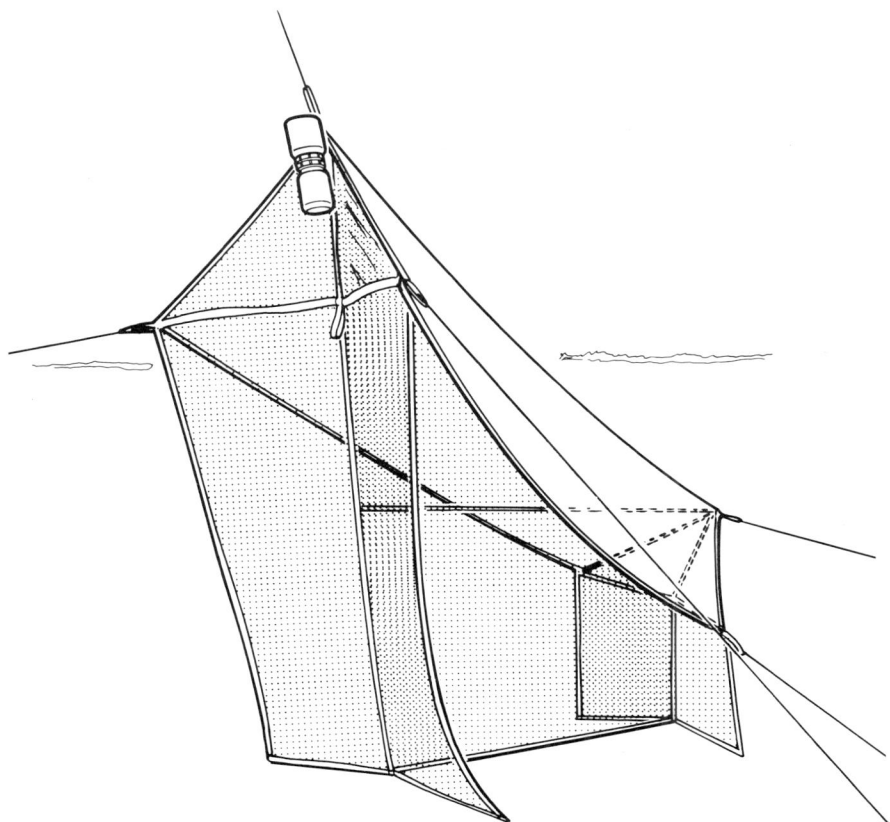

Fig. 12. Design and construction of a Malaise trap.

Fig. 13. A Malaise trap operating in the field.

Fig. 14. The collecting chamber of a Malaise trap.

Fig. 15. Another design of Malaise trap shown operating in the field.

Lepidopterists prefer not to use Malaise traps, partly because butterflies and other diurnals tend to get damaged in the trap but mainly because nocturnal Lepidoptera are caught so much more readily by light traps.

Many modifications of the original design by Dr. Malaise have been made (see "References," p. 172, and Figs. 12-15), but all the various designs have four things in common: one or more openings into which insects may wander, vertical walls that obstruct and restrict their passage, a roof that slopes upward toward the vertical walls, and a trapping device at the highest point on the roof that permits insects to enter easily but not to leave.

The precise shape of the Malaise trap does not seem to be very critical, but in practice several principles have emerged: the volume of catch varies with the size of the opening, especially the one on the downwind side, but openings facing other directions do not affect the size of catch because most insects fly upwind; the walls may be of dark material but it is most important that the roof is white in order to attract insects upward; and the exact placement of the trap is the most critical factor in determining the size of catch.

Because the trap is passive and has no attracting powers (except possibly for Tabanidae), place it across the flight path of the maximum number of insects. Experience and observation are the best guides. Avoid windswept open places and deep shade. In forested areas, traps collect the greatest number of insects when they are placed across woodland paths, small forest streams, and the sides of clearings that are sheltered from the wind. In open country, the most effective location for traps is in small gullys running transversely to the wind, in the lee of small patches of shrubs or trees, or in any other sheltered spots.

Because the Malaise trap is a large and conspicuous object, it may be stolen or destroyed by vandals. When you choose a site, keep these factors in mind. Avoid cow pastures as a possible site, because cattle like to rub against Malaise traps and they soon destroy them. Maintenance of a trap is simple: check guy ropes occasionally for tightness and look inside the trap, especially the entrance to the killing bottle, for spider webs, which must be removed and the spiders captured and killed, if possible. Change the collecting bottle at appropriate intervals, depending on the concentration of the killing agent you are using, if any, and the amount of insect activity.

If you can check the trap several times a day, you do not need a killing or preserving agent. You may be able to select the specimens that you want alive and to release the unwanted ones. For specimens that do not need to be kept dry, 95% ethyl alcohol is the best killing and preserving agent; it can be left in the Malaise trap until the alcohol is almost filled with insects. The length of time it takes to fill the trap depends on the size of the catch, the temperature, and the size of the bottle. The length of time varies from a few hours to several weeks, but 2 to 4 days is average. If you want a dry catch, potassium cyanide may be used as the killing agent, but it tends to dehydrate some specimens and to stain others. Therefore, the bottle should be changed often, at least daily in cool weather and several times a day in hot weather.

Dichlorvos is a more satisfactory killing agent for dry catches, because it stupefies the insect first and then slowly kills it. Thus the catches are not so affected by dehydration. If the catch is protected from the heat of the sun, the trap does not need as much attention as one equipped with cyanide. Other killing agents are chloroform and ethyl acetate, but the vapors of these organic solvents may damage the plastic of the trapping head and therefore they should be tested before you use them.

Window traps A window trap (Figs. 16 and 17) is effective for capturing crepuscular insects, especially small Coleoptera. This trap consists of a piece of clear glass or plastic suspended over a trail, log, or other habitat, with a shallow trough containing ethylene glycol beneath it. Most beetles drop when they hit an obstruction and then they are caught in the ethylene glycol.

Tent-window traps This trap (Fig. 18) is a rectangular tent 1.8 m high, 1.8 m long, and 1.4 m wide (6 × 6 × 4-1/2 ft). The ceiling and two sides are of black vinyl plastic sheeting, the front is clear vinyl plastic, and the back is open. The four corners are supported by vertical poles about 2 m (6-1/2 ft) long of either 2.5 cm (1 in.) dowel or metal pipe. A fifth pole of the same length supports the transparent front and is affixed horizontally between the upper ends of the two front poles. The trap is held rigid by four guy ropes attached to the upper ends of the corner poles and stretched diagonally to an anchor in the ground. Because the trap is based on the principle that insects usually fly toward light (for example, a window in a room), the trap should be located accordingly on the edge of woods, bushes, or trees in order to provide shade behind the trap. Wind may also be a problem with traps of this design. Set out the trap so that the wind is against a closed side, or preferably a closed corner. For best results, visit the trap every hour or two during the day. The trapped insects may be collected by hand with the use of an aspirator or a small net, or by putting them directly into a killing bottle.

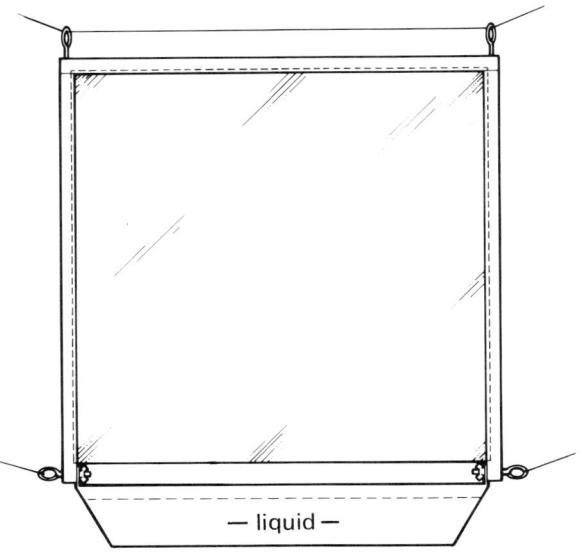

Fig. 16. Design and construction of a window trap.

Fig. 17. A window trap operating in the field.

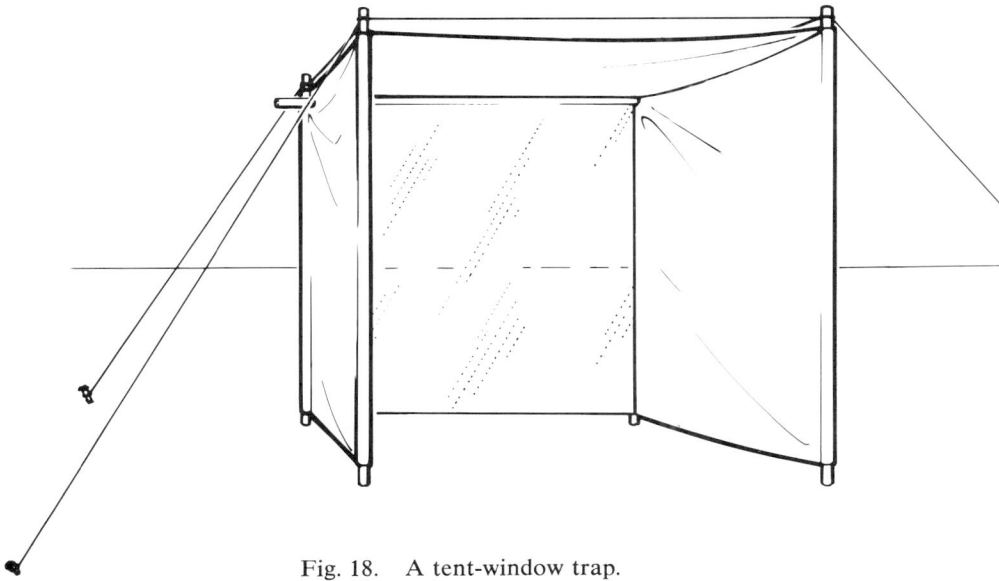

Fig. 18. A tent-window trap.

Visual-attraction traps This trap (Fig. 19) was originally designed to attract and capture mosquitoes by taking advantage of their response to a moving pattern of black and white stripes on a cylinder. Most groups of Diptera as well as some groups of other orders of insects can be captured by this type of trap. The insects may be collected dry or in a preservative.

Pan traps Ordinary aluminum cake pans (Fig. 20), about 23 cm (9 in.) square and 4 cm (1-1/2 in.) deep, are commonly used for pan traps, but almost any suitable size of container will do; in locations where someone may steal the pans, use sheets of plastic or aluminum foil. Set the pans in the ground with their tops flush with the surface. Fill them with water and add a small amount of detergent, which acts as a wetting agent, and 1–3% Formalin, which acts as a killing and preserving agent. If evaporation is a problem, as in hot or windy locations, or if the traps are left untended for long periods, up to 90% of the water may be replaced by ethylene glycol, which does not evaporate. The traps are best tended every 1 to 3 days; after 3 days, osmotic action starts to occur and it causes abdomens and other soft parts of the insects to swell.

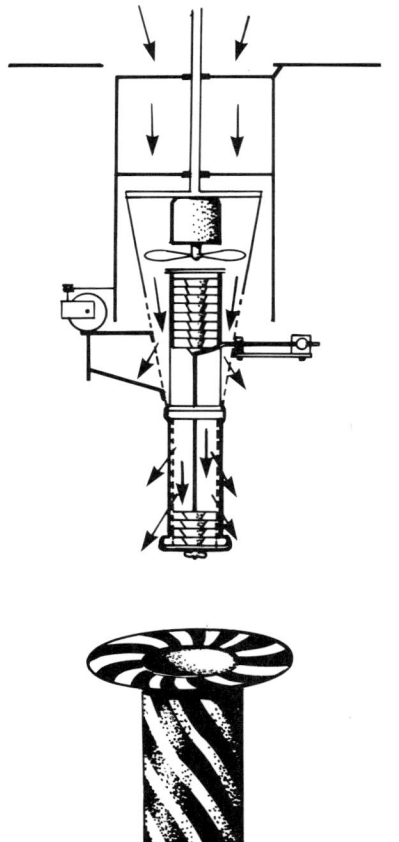

Fig. 19. A visual-attraction trap.

Fig. 20. A pan trap in operation.

The traps function as simple pitfalls for wingless creatures such as beetles, larvae, Collembola, and spiders, but they also catch large numbers of flying insects that land on the water surface or are carried there by wind currents. The traps are especially effective in cool windswept areas, where wind speeds are highest near the surface of the ground and insect activity is confined to the first few centimetres (inches) above ground level. Thus they function very efficiently in arctic tundra and grasslands. In woodland areas, where there is only slight wind movement near the ground, you may increase the catches by placing sheets of clear plastic vertically above the center of the trap in order to intercept flying insects.

The easiest way to service the pans is to pour the contents through a fine strainer, about 20 meshes/cm (50 meshes/in.), into a bucket; then replace the pan and pour the contents of the bucket back into the pan and add enough water to replace losses from evaporation. To wash the Formalin, detergent, and ethylene glycol off the specimens, empty the contents of the strainer into a collecting vessel full of water.

Baits The use of natural and artificial or chemical baits is a productive method of collecting insects and other arthropods. To use natural baits, supply an abundance of food or breeding material either in traps or in marked areas. Baits commonly used are dung, feathers, raw skin, hair, bones, meat, carrion, logs, and boughs. Visit your baits regularly during the day and night, and collect your specimens by various methods, depending on the bait and its condition. Berlese funnels, sifters, or other means of sieving are useful, but some species attracted to certain baits require handpicking, beating, or netting. A headlight (p. 42) is best for night collecting if you are using the last three methods.

In certain seasons, fresh cuttings of various species of trees attract many of the bark- and wood-inhabiting insects as well as their parasites and predators. Visit the trees at frequent intervals of the day and night during the collecting season and handpick the insects or collect them by beating. Because many of the insects attracted are active, fast flyers, approach the site cautiously and collect carefully, otherwise many specimens will escape. Cage logs, branches, and twigs from cut trees after they have been subjected to insect attack for some time. Keep your cages under natural conditions or the woody material will become dry and many of the insects, especially the immature ones, will dehydrate and die. Natural attractants such as aphid honeydew, "bleeding" trees, and so on attract many insects; an alert collector makes frequent visits to these sites during the day and night.

Artificial or chemical baits are used in traps by painting or spraying them on trees. Sugaring for Lepidoptera (*see* "Lepidoptera," p. 138) is a well-known and widely used method. Malt, yeast, sugar, molasses, and oatmeal are also commonly used as baits. Malt and yeast or molasses diluted with water used in a pitfall trap (Fig. 21) is attractive to many beetles and other insects. To prepare malt bait, mix 71 mL (1/8 pt) of malt and one package of dry yeast into 4.5 litres (1 gal) of water. Allow the mixture to ferment for 2 or 3 days before using it. Fill the containers 13–25 mm (1/2–1 in.) deep with bait. Strain the catch through cheesecloth and wash it in alcohol. Insects such as camel crickets, which swallow the solution, have to be eviscerated or slit through the pleurites into the crop and placed in 40–50% alcohol. Change the alcohol several times until the solution is clear. If specimens containing molasses are pinned, they become discolored and the pin corrodes. Oatmeal trails through wooded areas or other suitable habitats attract certain ground-inhabiting beetles. Scatter the oatmeal along a path or on other bare areas at dusk. Visit the bait at intervals after dark and pick up specimens by hand or with forceps. Carbon dioxide as an attractant for biting flies, especially tabanids, has been used widely in Malaise and other traps. Traps can be adapted to hold containers of dry ice, or CO_2 can be supplied to the trap from a cylinder.

Boards coated with tanglefoot and placed in suitable situations often capture insects in large numbers. This method is not desirable for obtaining specimens for taxonomic study, because the specimens cannot always be identified, owing to damage by the tanglefoot. More insects are captured on yellow boards than on boards of other colors.

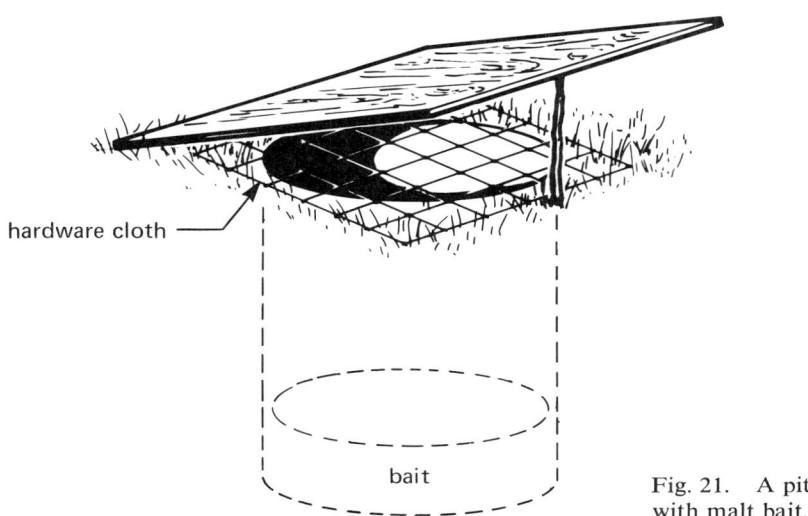

Fig. 21. A pitfall trap with malt bait.

Sweeping

Sweeping is the most productive method of collecting insects that fly among or crawl on herbage. Such insects can be captured in large numbers. However, sweeping has definite limitations, sometimes not fully appreciated. Its limitations in a faunal survey are that it captures free-living insects only; miners, borers, and leafrollers escape. Moreover, if the vegetation is at all dense, insects that live down near its roots escape capture. Where host-plant data are required, supplement sweeping by searching and observation, unless the plant understory grows as a pure stand. As a sampling method sweeping produces general indications, rather than accurate data, because of differences in the sweeping techniques of different people and even of the one person under varying conditions, and because of differences in the positions of insects on the vegetation at different times and under various weather conditions. Nevertheless, sweeping is the most efficient and one of the most rapid and simplest methods of collecting insects in large numbers from vegetation.

The purpose of sweeping is to dislodge insects from the vegetation by means of a net. Sweep the herbage, the flowers, or the foliage of trees with rapid sideways strokes of the net. Too-slow strokes allow the insects to escape from the net and upward or downward strokes normally produce fewer captives. To reduce damage to fragile specimens, examine the contents of the net and remove the insects after every few sweeps. The specimens may be picked from the net with your fingers, a fine-pointed forceps, or an aspirator. Or, you can group all your catch in the end of the net and trap it there by squeezing the net just above the group with your hand. Insert the part of the net containing the catch into a wide-mouth killing jar and hold your hand over the end until all the insects are inactivated or dead. By this method, you can examine the catch later. Do not sweep when the vegetation is wet, because you may damage most of the insects.

The sweep net, ring, and bag must be strong and durable to endure continuous brushing through vegetation. The rings described previously for netting (*see* "General-Purpose Nets," p. 11) are suitable. The most convenient length of handle is about 1 m (3 ft). The bag should be as deep as the diameter of the ring; if it is too deep, it is too hard to pick the insects out of the bottom. A suitable depth is 50–60 cm (20–24 in.) with a 38 cm (15 in.) ring. A small net with a ring 20–25 cm (8–10 in.) in diam is very useful for sweeping individual plants growing in mixed vegetation. The material of a sweep net bag must be tough enough to withstand sweeping through dense foliage without snagging or ripping, and the mesh must be fine enough to retain minute insects. Low resistance to air is not so important as in an aerial net. For these reasons the bag does not have to be made of netting; any strong, light cloth, such as tough scrim or nylon, is suitable.

Beating

Beating is one of the most productive methods of collecting insects, mites, and spiders from the foliage of trees and shrubs. It is particularly useful for collecting beetles, phytophagous insect larvae and mites, and predaceous mites and spiders. But it is not fully reliable as a method for obtaining host-plant data, because some arthropods not usually associated with the plant that is being beaten may alight on it or on the beating cloth or may crawl onto it from the herbage below. Reasonably reliable host-plant data for plant-feeding forms can be obtained by beating only when the host-plant associations of the insects and mites captured can be checked by rearing the insect or mite on the species of plants from which it was beaten, or when there is strong evidence shown by the regular occurrence of a species of insect or mite only on a particular species of plant. However, be careful not to assume that every arthropod found while beating has been beaten from its host plant. Also, it is difficult to be sure that the beating sheet is clean when you start beating a different species of plant. Many species of insects can be collected by beating dead branches or clumps of dead leaves.

The principle of beating is to hit a branch of a tree or shrub hard enough with a heavy stick that the arthropods fall on the tray or sheet below, where they are easy to see and capture. Though beating is best for collecting larvae, many active free-living and flying insects may be found, particularly if the weather has made them lethargic. Strike the branch with a downward stroke; a sideways stroke may cause some specimens to fall beside but not on the sheet. Some collectors give a branch two sharp blows in rapid succession, on the theory that the first stroke loosens the arthropods hold and the second dislodges them. Beating at night with the aid of a headlight is particularly effective for collecting beetles and other insects. Do not neglect dead and dying trees and branches because many nocturnal wood borers are active there.

Use a strong stick for beating. The cloth onto which the insects fall may be a sheet, a net, or an umbrella in your hand. A beating sheet or net is preferable to a cloth laid on the ground because it prevents the capture of insects that crawl from the herbage. If you are collecting larvae, withdraw the sheet from strong sunlight after you beat the branch, because many larvae die quickly when exposed to heat.

The beating sheet shown in Figs. 22 and 23 is made of nylon shaker cloth, about 1 × 1 m (3 × 3 ft), with corner pockets sewn on both sides, making it reversible. The centerpiece is welded and made from electrical conduit as are the four rods, which are smaller in diameter so that they readily slide into the centerpiece. Holes and snap catches are useful for adjusting the tension on the sheet, particularly on humid or wet days. The terminal ends of the rods have a stiff wire adapter to hold each rod firmly in the corner pockets. Rods of 16 mm (5/8 in.) hardwood dowel can be substituted for the electrical conduit. This type of beating sheet is readily collapsible and is easy to pack.

Instead of a beating sheet and stick, a screened beating tray (Fig. 24) is more effective for beating small arthropods, particularly mites, from foliage. Rather than striking the branches with a stick, strike the branches two or three times against the upper screened surface of a white enamel or plastic tray. The specimens fall through the screen to the surface of the tray and the screen prevents excessive fragments of leaves and debris from cluttering the tray. Immediately after you finish beating, remove the screen and examine the white surface of the tray for tiny moving specimens, which can be collected either with an aspirator (see "Aspirator," p. 43) or with a fine artist's brush moistened with alcohol (see p. 162). White photo-developing trays, 20 cm wide, 25 cm long, and 5 cm deep (8 × 10 × 2 in.), made of high-impact plastic are suitable for this purpose. Bend the screen of 6 mm (1/4 in.) mesh over the sides of the tray so that it will not dislodge during beating.

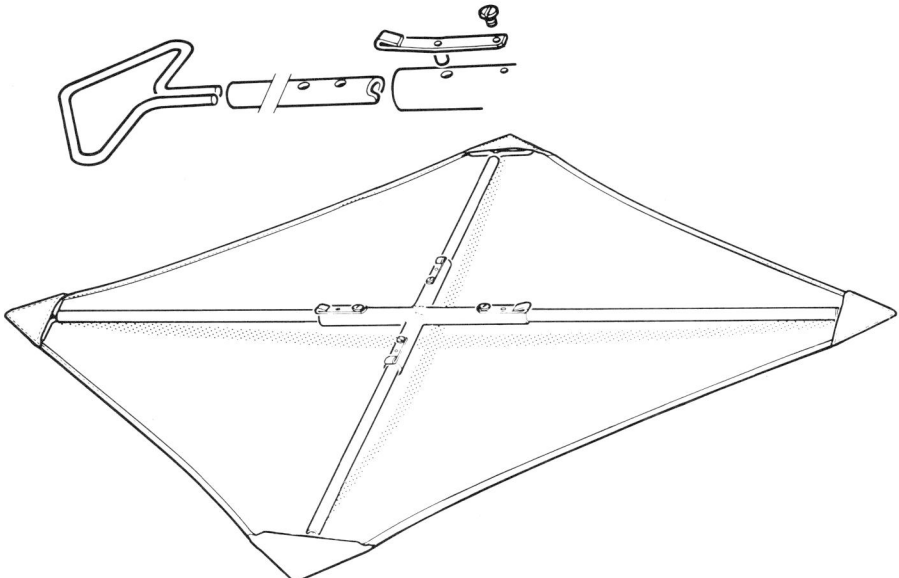

Fig. 22. Design and construction of a beating sheet.

Fig. 23. A beating sheet in use in the field.

Fig. 24. A screened beating tray collecting mites from foliage.

Or, you can use a beating funnel (Fig. 25), which is a fairly large funnel about 30 cm (12 in.) in diam at the wide upper end, tapering to a short 19 mm (3/4 in.) in diam metal tubular extension at the lower end. A short piece of flexible rubber sleeving fits over the tube, and a 19 mm (3/4 in.) in diam collecting vial, half-filled with alcohol, can then be fitted over the rubber sleeving. Place a 6 mm (1/4 in.) mesh screen over the wide upper end of the funnel when you are using it, and strike the foliage against this screened surface in the same way as with the beating tray described previously, but with somewhat different results. The specimens and organic particles fall through the screen and collect in the vial; the method is fast but the specimen content of the collection is uncertain and sometimes cluttered with debris.

A power-driven mite-brushing machine (Fig. 26) has been used for quantitatively assessing populations of spider mites infesting the foliage of trees. This method has not been assessed for qualitatively sampling the various other kinds of mites and small arthropods that live in trees.

A suction or vacuum machine (Fig. 27) is a power-driven device used for quantitative collecting of insects, mites, and spiders. It is efficient for extracting the arthropods with harder bodies from open fields, rough grasslands or herbage, dense shrubbery, and heavily forested areas where the long flexible rubberized hose can be utilized to collect the high canopy. Several designs of suction machines are available and can be adapted for specialized collecting. A backpack model is available that allows greater mobility and leaves your hands free.

Fig. 25. A beating funnel.

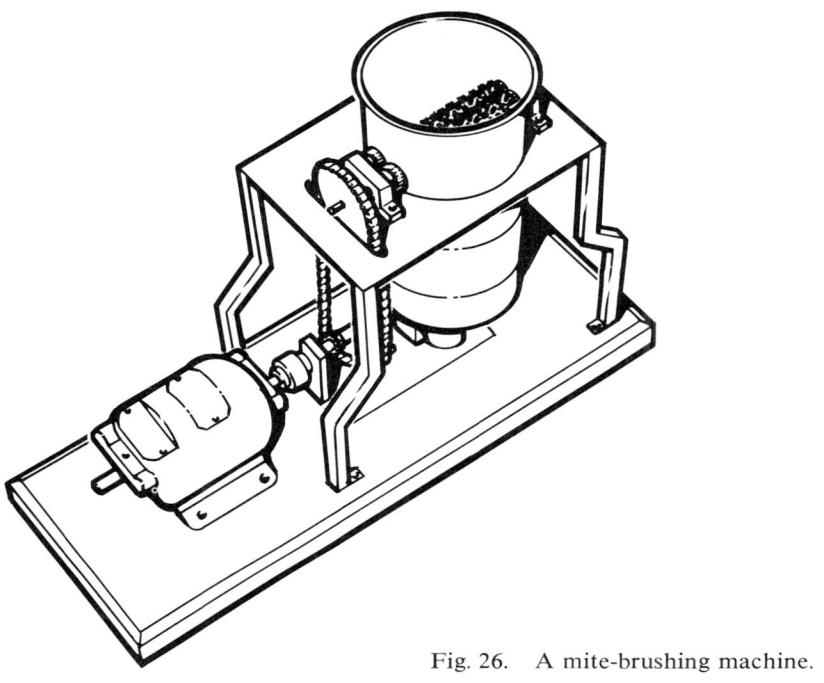

Fig. 26. A mite-brushing machine.

Fig. 27. A vacuum machine sampling spider populations.

Emergence cages and separators

Emergence cages may be used to collect flying insects that emerge from soil, debris, water, or low vegetation. These insects are usually difficult to collect in numbers by others means. The principle is to cover with a closed cage an area of ground or water in which the insects are developing. After they emerge, the insects can be picked or aspirated off the walls or trapped in a container at the apex of the cage.

A type of cage used to collect adults of flying insects whose early stages are aquatic is shown in Fig. 28. Set it up over a part of the bed of a stream or lake where the early stages occur and visit it regularly until all the adults emerge. They will be found resting on the walls and roof of the cage. To obtain adults of species that develop in deeper water, place the cage on a raft that has an opening corresponding to the opening in the bottom of the cage.

A cage for trapping insects is based on the principle that flying insects tend to move upward. The design of the cage may be modified, but it is usually in the form of a pyramid with an opening at the apex. It may be made of cheesecloth, wire screening, or other suitable material, but preferably should be dark and fairly opaque. It should cover a suitable area of ground or vegetation. The insects make their way upward and pass through the opening into a killing bottle or other container. A simple cage of this type is shown in Fig. 29.

The larger the emergence cage (that is, the greater the area it covers), the larger the number of insects you will capture. A square cage that is 1 m (3 ft) at its base is a suitable size, but the larger it is, the better. After the cage has been set up for a day or so, insects that happen to be in the vegetation may be found in the cage. But do not conclude that all insects found in the cage in the first few days have developed in the area covered by the cage.

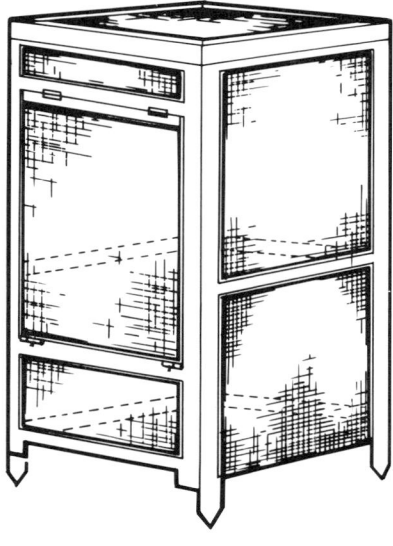

Fig. 28. A black fly cage may be used as an emergence cage for aquatic insects.

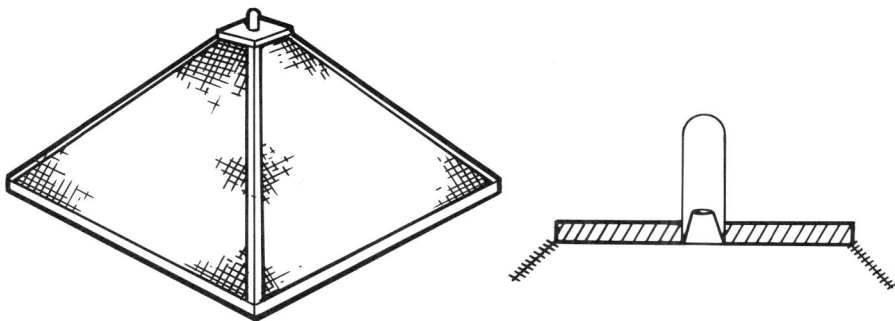

Fig. 29. A simple emergence cage showing detail of the top of the cage with a killing bottle.

A separator is similar in principle to an emergence cage: it is based on the principle that diurnal insects tend to move toward a light in a dark container. Its chief use is to separate insects from the mass of debris that collects in a sweep net, but it can also be used as an emergence cage by putting in it lumps of matted vegetation, rotten wood, dead or dying branches, bolts of tree trunks, or other materials in which insects are developing.

A separator is a box of any convenient size, with a tight-fitting lid. At one side near the top there is an opening into which a tube or killing bottle fits (Fig. 30). After the debris has been put in the box, the insects make their way upward toward the light, and they congregate in the tube. A ramp inside the box, from the floor to the tube, enables the insects to reach the tube more easily. Never leave a separator in the sun, because internal condensation and overheating readily occur. Because of condensation, particularly in hot weather, or if there is soft, moist vegetation in the separator, or an accumulation of grasshoppers, spiders, and other animals that may damage the insects or each other, change the tube often while the separator is in use. Some insects do not move upward and toward the light; these can be found later by sorting through the debris remaining in the box.

Separators or emergence cages range from elaborate rearing chambers with interior temperature and humidity controls to simple paper sacks or cardboard ice-cream containers with a vial inserted in one end. The type and style used depend on the needs of the collector.

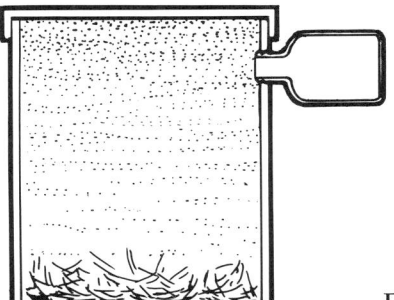

Fig. 30. A separator box.

41

Searching

This method of collecting is not used as often as it deserves, possibly because it is somewhat less productive of species and specimens than are certain other collecting methods. However, searching for and collecting individual specimens by hand is by far the most accurate and reliable way of finding out what species are associated with particular plants or habitats, and results in the acquisition of valuable biological data at the same time. Do not rely too much on your memory, but make notes while you are in the field. To search effectively for particular species, you need to be familiar with the soils and flora, the plant structure, and the seeding characteristics. A good field naturalist is the best collector.

Searching is the most profitable method of collecting series of arthropods whose habits are such that they are difficult to collect by other means, for example, insects that mine or roll leaves (see "Collecting Leafmining Insects," p. 46); insects and mites that cause and inhabit galls; arthropods that live among the roots of grasses and other plants; arthropods that inhabit crevices or are found under bark (see "Barking," p. 42); insects that are relatively immobile or that are firmly affixed to some object; and arthropods that are found in beach drift or flood debris. The species found are often of particular interest or rarity, because relatively little attention is paid to this method of collecting. Free-living and flying species may be picked off vegetation and stones, particularly on dull days.

Searching at night with the aid of a head lamp or in the twilight before dawn is an excellent method of collecting nocturnal insects, especially Coleoptera. Dead or dying trees, rotten stumps or logs, fungi, beaches, stream banks, wet spots such as springs or seepage areas, and sand dunes are productive and interesting locations for night and twilight collecting.

You can pick up the insects that you find with fine-pointed forceps or your fingers. A small moistened artist's brush is useful for picking up small insects and arachnids. However, the most convenient way of picking up these arthropods is with an aspirator or a collecting bottle.

Barking This method of collecting certain insects is seldom used, but it is particularly productive for certain species of Heteroptera and many Coleoptera, which are found only in or under bark. The bark of trees recently cut down for pulp or lumber and the bark of dead trees still standing are most productive. The bark on stumps supports large populations of insects, but logs or trees where the bark is just beginning to separate from the wood are the best for this purpose. Most of the insects are found directly under the bark and in tunnels and crevices of the bark. In the latter case they must be jarred loose. This is best accomplished by striking each piece of bark against a hard object until the insects fall onto the beating sheet. You can pick up these insects by hand or with an aspirator and place them in killing bottles.

Mammal nests A number of insects live in ground nests and the burrows of various mammals. The best way of collecting them is to dig out the nest and search it and the soil around it (some species are in the soil under or around the nest, but not in the nest itself). Because it is often very difficult to dig out most mammal nests, other methods have been developed. A very useful method, especially in high elevations where the burrows are often in rocky terrain, is to push a small package of hay or dry grass with some strong-smelling cheese or old meat wrapped in a fine wire screen and fixed on a long and strong wire deep into the gallery. The cheese or meat attracts insects, especially beetles, which congregate in the hay. In a few days pull out the trap with the insects in it. One disadvantage of this method is that the trap is sometimes destroyed by the mammal. However, because you usually set out several traps, the method is quite effective.

Some insects leave the nest, enter the galleries, and congregate around the entrance and in the front part of the gallery behind the entrance. These can be collected by pulling out the soil from the entrance and the gallery. This method is especially suitable in the spring for collecting in the burrows of groundhogs, ground squirrels, and prairie dogs.

Insects living in nests on or above the ground (for example, those of beavers, muskrats, and wood rats) are easy to collect by sifting all the nest material. Species living in muskrat nests may also be collected by treading the central part of the nest pile into the water and then collecting the beetles from the surface of the water.

Aspirator

An aspirator is a simple suction apparatus that is used for picking up numbers of insects and arachnids or for selecting individual specimens out of a large number or off a plant. There are several designs (Figs. 31–33), but the instrument used most widely is a vial of glass or, preferably, of transparent plastic, with a close-fitting rubber stopper (Fig. 31). Two tubes pass through the stopper. A rubber tube is attached to the outer end of one of these tubes. Through this, you suck with your mouth, squeeze a suction bulb, or use some suction-producing apparatus such as a modified hair drier or vacuum cleaner. The inner end of this tube is covered with a fine cloth or screen to prevent insects from entering. The second tube (Fig. 32) is open at both ends and projects into the container. To use the aspirator, place the outer end of the aspirator, the intake tube, near an insect or arachnid and apply suction sharply through the suction tube. This suction creates a partial vacuum in the container and draws the arthropod up through the intake tube. Be careful to avoid contacting parasitic arthropods or other arthropod-borne pathogens when you are operating the aspirator by mouth. Aspirator filters do not entirely prevent this danger, so it is best to replace oral suction with bulb suction when you are collecting ectoparasites and soil arthropods.

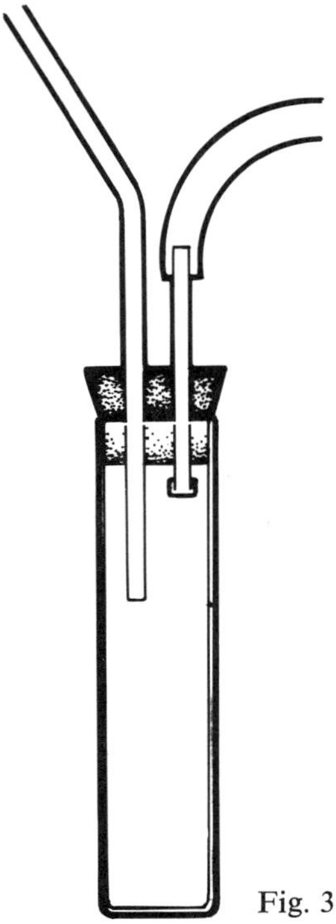

Fig. 31. An aspirator.

Fig. 32. An aspirator.

Fig. 33. An aspirator.

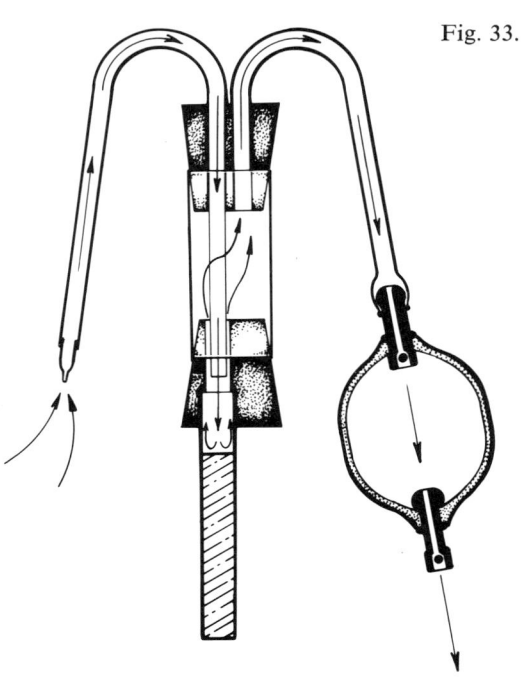

44

The use of a bulb or Singer aspirator is highly recommended for collecting mites, small insects, and spiders, particularly from plants, because it collects directly from the habitat into the preservative and avoids the problem of handling delicate specimens. Also you can make separate collections in rapid succession without any danger of mixing the specimens. The intake and suction tubes are made of clear, flexible vinyl tubing with an inner diam of 4 mm, and their lengths can be determined by your needs. The intake tube is tipped with an interchangeable nozzle made of a short length of glass tubing with one end drawn out to give the desired aperture size. Two sizes of aperture, 1.5 and 3 mm in diam, are useful for general collecting. Two No. 2 rubber stoppers are necessary: one has two holes to allow for a tight fit with the intake and suction tubes; the other has one hole, with a smaller interior diameter and a larger exterior diameter (Fig. 33). The narrower portion of this hole must be 1 mm larger than the outer diameter of the intake tube to allow for free movement of air through the system. The wider portion of the same hole is 1 mm smaller than the outer diameter of the collecting vial in order to grip this vial securely while it is in use. The body tube is made of metal or heavy-duty glass or transparent plastic with an inner diam of 18 mm; it should not extend over the part of the lower rubber stopper that grips the collecting vial. The collecting vials are glass, 8–9 cm (3–3-1/2 in.) long, with an inner diam of 6 mm (1/4 in.); they should contain 90% alcohol because lower concentrations present a surface tension that may trap specimens. A pipe cleaner is suitable for cleaning and drying parts of this aspirator when arthropods begin to stick along the intake tube or when condensation forms in the body tube.

Condensation tends to take place inside an aspirator. To prevent this, it is advisable to insert a piece of dry blotting paper. The stopper should be tight fitting; otherwise small insects may crawl up between it and the wall of the container and get crushed. Do not put a killing agent in the aspirator while it is in use, partly because it is ineffective as its fumes are removed by the suction and partly because of the danger of the operator inhaling poisonous fumes. The insects in the container may be killed by inverting the container into an open killing bottle; by putting the whole container, with the stopper removed, into a large killing bottle; or by replacing one of the stoppers with one that contains a killing agent (*see* p. 73).

A collecting bottle (Fig. 34) is actually an aspirator without suction. It consists of a jar or vial of convenient size with a rubber stopper through which a glass tube passes. This tube projects into the jar, to prevent the insects from escaping or falling out if the jar is inverted accidentally. Pick up the insects with your fingers or forceps and drop them into the tube. Instead of the tube you may use a thin-walled plastic tube with its outer end cut off diagonally to make a small scoop (Fig. 35) with which the insects can be gathered up. An advantage of a collecting bottle over an aspirator is that it can contain a killing agent; then it is a killing bottle with a tube through the stopper.

Fig. 34.
A collecting bottle being used as a killing bottle.

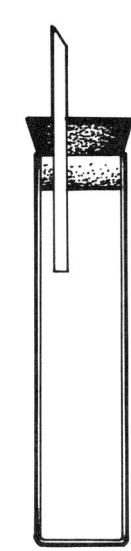

Fig. 35.
A quill collecting bottle.

Collecting leafmining insects

Many species of larvae of Lepidoptera, Diptera, Hymenoptera, and Coleoptera feed within the leaf, needle, petiole, and cuticle of plants; these are commonly referred to as leafminers. Most of these insects are small (or micro) and are considered separately from the burrowing and boring insects described in the following chapter. Miners may be found on any plant and you should always collect enough of the host plant for future identification if necessary. Because these insects are collected in the immature stage, the material has to be reared to obtain adults. This can be done successfully by placing the material in a 227 mL (8 oz) screw-cap jar or other similar container. Spread the foliage throughout several jars to avoid crowding and the risk of losing the collection. It is useful to press samples of the mined leaves and to mount these in a scrapbook. Later, when the adults emerge and have been identified, the name of the miner can be associated with the host plant.

Most collectors are interested in only one order, and with experience it is possible to differentiate among the orders by the characteristics of the mine and the habits of the larvae. Although the habits and mines vary within the order or group, you may find the following information helpful. Mines devoid of frass or those that the larvae have left to continue feeding elsewhere are lepidopterous. Those mines that show primary and secondary feeding tracks or that have their frass lying on one side and then on the other in a linear mine are dipterous. Also, the frass of Diptera larvae often appears to be liquid. The frass of hymenopterous miners is usually scattered irregularly in the mine. A coleopterous mine usually has a spot of shiny secretion covering the egg at the entrance of the mine.

Collecting and extracting arthropods from debris

Many small arthropods are found in humus, dead and decaying vegetation, beach drift, flood debris, the nests of birds and mammals, and similar litter and debris. Except those whose adults are winged and active and may be obtained with an emergence cage or a separator, these forms are rarely or never obtained by the collecting methods already described. They can be collected by searching and handpicking or by forcing them to leave a sample of the debris by altering its character so that conditions in it become unfavorable to them.

Berlese funnel Ever since it was originally designed by Antonio Berlese early in this century, the modified Berlese funnel has been the most practical device for extracting the small forms that are the most common arthropod life in organic soils and leaf litter. It can also be used successfully for extracting insects from various other habitats, such as loose bark, rotting wood, bracket fungi, mosses, and flowers; other specific parts of higher plants; stored food products, manure, and freshly killed vertebrate animal hosts; and the nests of birds, mammals, and social insects. Although the Berlese funnel is used especially for extracting Acari and Collembola, it is effective also for Isopoda, Araneae, Opiliones, Pseudoscorpionida, Myriapoda, Protura, Diplura, Thysanura, Psocoptera, Siphonaptera, and ground-dwelling and sometimes flightless forms of Thysanoptera, Hemiptera, Coleoptera, Diptera, and Hymenoptera. Yields of holometabolous insect larvae may be good if the larvae are mobile, as in the case of the legged Coleoptera and the apodous (footless) Siphonaptera and some Diptera.

The essential components of a Berlese apparatus (Fig. 36) are a metal sample holder made of wire mesh or screening on the bottom; a metal or plastic funnel in which the sample holder is placed; a wooden grid or baffle, or metal screen, placed in the funnel below the sample holder, to trap bits of detritus falling from the sample; a collecting vessel that attaches to the narrow lower end of the funnel and usually contains a liquid preservative; a lid covering the sample holder and the funnel that encloses the source of the extracting stimulant; the extracting stimulant (a light bulb or heating element for desiccation, or a volatile chemical repellent) placed under the lid and above the sample holder; and a frame or stand that holds and stabilizes the funnel.

To operate a Berlese funnel, place a sample of the habitat on the screen or wrap the sample in one or two thicknesses of gauze or cheesecloth and put it on the sample holder in the funnel, and supply a source of heat and desiccation (an incandescent light bulb, electric resistance wire, or a hot-water jacket) or a chemical repellent (a naphthalene and paradichlorobenzene mixture or chloropicrin) above the sample. The arthropods react to the heat or repellent by moving downward, away from the heat, and deeper into the sample. Finally they fall through the screen at the bottom of the sample and collect in a vial, a screw-top jar, or a stiff plastic sealable bag suspended below the narrow end of the funnel. To prevent excessive amounts of detritus

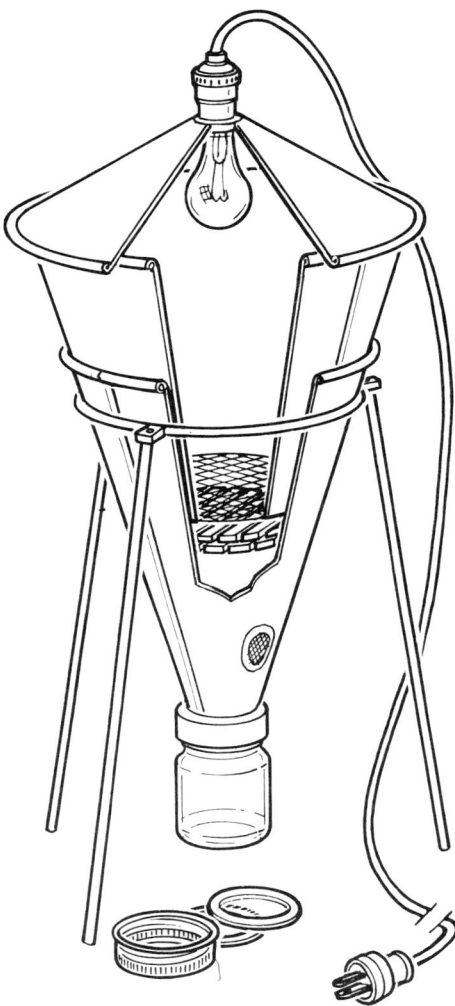

Fig. 36. A Berlese funnel.

from being brought down the funnel by the activity of the specimens, place a grid, screen, or baffle (Fig. 37) between the sample holder and the collection bottle. Experience has shown that these filtering devices produce a cleaner sample of organisms, which greatly aids in the final sorting and mounting of specimens from the sample. However, these barriers must not be so elaborate that they prevent the arthropods from moving down the funnel.

A funnel with a light bulb as the repellent source is the most common type. This is based on the principle that most arthropods in the substrate samples prefer dark, moist conditions and move away from both the light and the source of drying of the sample. Select the wattage of light bulb best suited to the size and form of the funnel and the size and wetness of the sample.

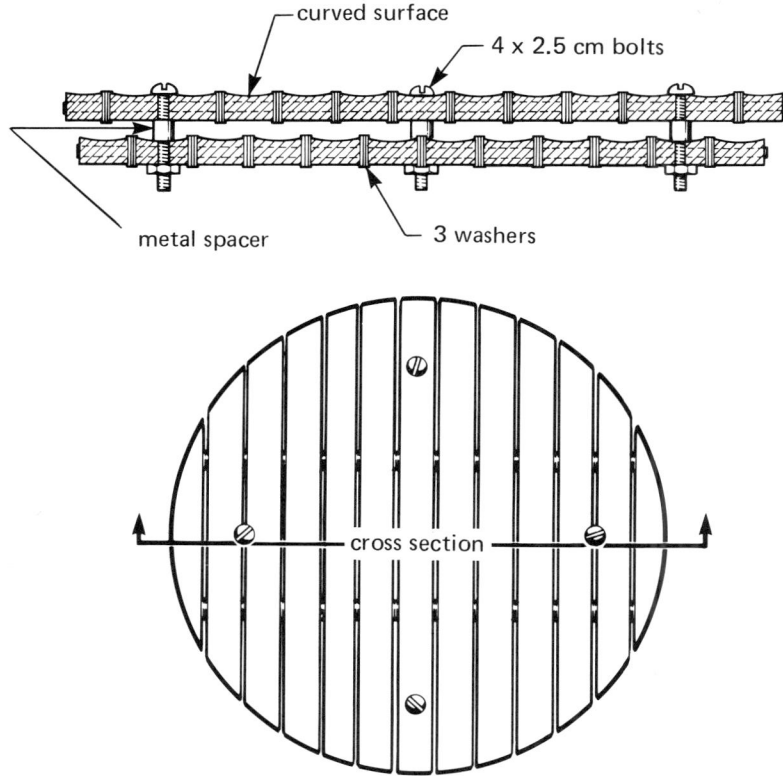

Fig. 37. Design and construction of a wooden baffle.

Bulbs of over 100 W may kill many of the light, sclerotized, slower moving arthropods, whereas bulbs of under 40 W take too long to produce satisfactory results. Generally, samples of about 4.5–9 litres (1 or 2 gal) in volume in a moderately large funnel of 30–38 cm (12–15 in.) upper diam with a light bulb of 60 or 100 W process well in about 2 days. Samples of this size usually yield members of species that occur even sparsely in the habitat.

A lid or hood that closes over the light bulb and the sample holder of the funnel is advantageous in several ways: the heating effect and desiccation gradient are more pronounced; organisms capable of jumping or flying out of the sample are prevented from doing so; contaminant organisms (particularly flying nocturnal insects that often carry mites) attracted to the light or to the odor of the sample in the funnel are excluded; and funnels being used outdoors are protected against rain or other falling objects.

Nonflying arthropods may crawl up the legs or sides of a unit and get into the funnel. Many funnels are not designed to eliminate these problems of contamination because they are used mostly inside laboratories. But a lid, tight-fitting parts, and fine-meshed netting covering the points of entry greatly minimize these problems outdoors. If you do not have a lid, cover the open top of the funnel and sample holder with one or two layers of gauze or other netting material to prevent contaminating the sample and organisms from escaping. The netting also shields samples from wind when funnels are operating outdoors.

If you do not have electricity, heat, or light, use chemical repellents with an accessary part for the funnel. Attach a fine-meshed (1–2 meshes/mm or 32–48 meshes/in.) metal-screened, shallow basket to the upper edge of the sample holder of the funnel, just below the lid. Sprinkle a couple of handfuls each of moth crystal flakes of naphthalene and of paradichlorobenzene in a thin layer in this basket, or apply a few drops of chloropicrin to a thin layer of cloth or cotton in the basket. Chloropicrin, better known as tear gas, has the advantage that only small quantities are needed and it acts fast, extracting most arthropods within 2 hr under favorable temperature conditions; its drawback is that it irritates the skin and mucous membranes if it is overused. Moth crystals are easier to use; they also obtain relatively fast results in 6–12 hr. Remember that, without an electrical source of heat, the reaction and activity of the arthropods in a funnel sample to a repellent depend upon the temperature conditions of the surroundings. If you are processing samples outdoors, therefore, you may obtain good results with chemical repellents in mild or warm weather, but poor, misleading results in cool or cold weather (particularly from samples run overnight in mountainous regions). In addition, for faster results and independence from electricity, chemical repellents have some other advantages: arthropods do not die because of succumbing too quickly to heat or desiccation, or because of becoming trapped in substrates such as heavier mineral soils as they dry out and set into a hard mass; arthropods are extracted about as quickly from wet samples as from dry ones; and the extraction is usually cleaner because of minimizing both the time of activity of the active components and the crumbling, powdering effect of desiccation of the inactive components (especially humus and soil).

Do not use chemical repellents in conjunction with a heating-desiccating method for extracting. Also, do not place chemical repellents directly on the surface of the sample, because this may kill some of the arthropods nearest the chemicals and also hinder formation of a clear-cut gradient of repellent concentration from top to bottom.

A separate sample holder, or hopper, may be incorporated into the design of an extracting funnel (Fig. 38). The sides of the sample holder should have the same taper as the funnel to allow for the tightest possible fit in the funnel and for the nesting of similar parts of several funnels together for transporting and sorting. Although it is not absolutely necessary and it adds weight and complexity to the extractor, a hopper helps obtain a cleaner and faster yield of arthropods. The hopper can be removed easily at the outset of processing a sample to clean off loose dirt that falls onto the segregating grid. Subsequently, you can speed up the rate of collecting by removing the hopper two or three times, taking out the dried upper layers of the sample, and turning or shaking the sample before placing the hopper

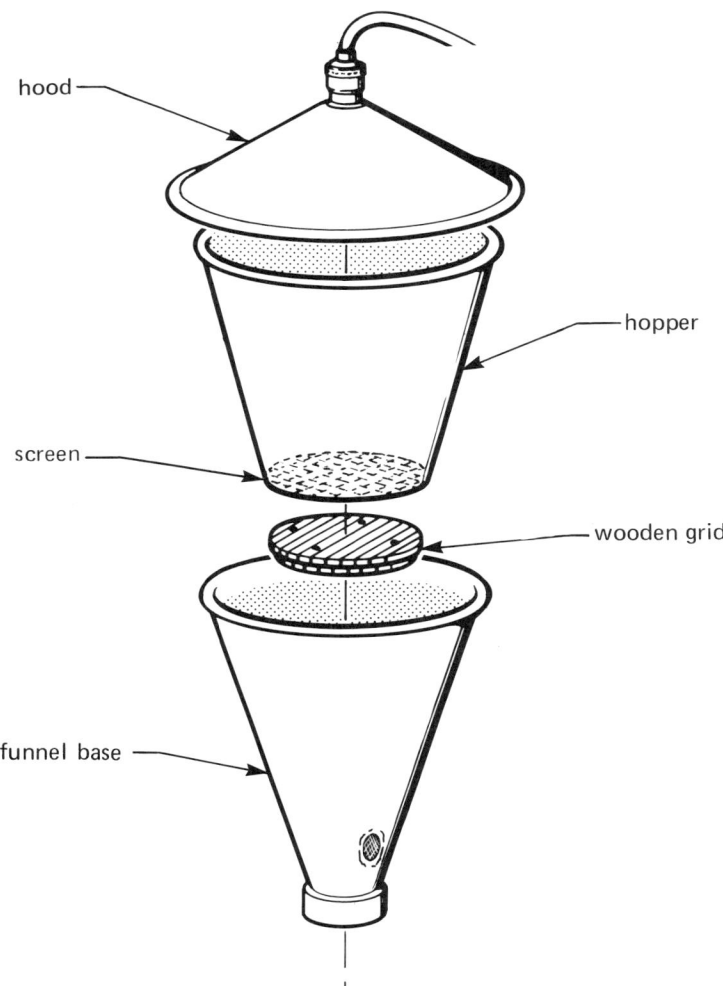

Fig. 38. A sample holder.

back in the funnel over the grid. Also, a ventilating space (called the Haarlov passage) between the lower edge of the sample holder and the funnel provides ventilation and speeds up the rate of desiccation within the funnel.

For qualitative work, simple Berlese funnels are usually adequate. For quantitative studies, modified designs have been made, such as the high-gradient funnel and the controlled-draft funnel, and the split-funnel extractor and the bowl extractor. These designs are more elaborate and are not described here, because they are not used in the field or on expeditions. However, the importance of air passages to the efficiency of the funnel should be considered when you are constructing a funnel. A lot of air, which can be utilized or controlled, occurs at four places in the apparatus: around the edge of the lid; between the lower edge of the sample holder and the funnel; at the top of the lid, if you make a damper outlet; and at or near the bottom of the funnel.

To operate an extracting funnel, invert the sample when you place it in the funnel. In this position the upper layers of the sample, which often contain the greatest number and variety of arthropods, are placed downward in the apparatus and the arthropods have to move a shorter distance and require less time to escape. The portability of funnels is of prime concern for field trips, particularly expeditions. Collapsible funnels of canvas or plastic sheeting are sometimes used (Fig. 39), but some arthropods become entrapped and die on the rough surface of cloth funnels, and plastic sheeting is delicate and difficult to maintain; moreover, these collapsible funnels are hard to support and stabilize. Simple mass-produced, pliable plastic funnels of 20 cm (8 in.) in diam may be used as the base for workable units. Larger sizes (30 cm, or 12 in.) are better but they cost too much. The other components used with a plastic funnel are shown in Fig. 38.

The stands (Fig. 40) for funnels can be made collapsible, like a tripodal oversized ring stand. A portable extractor has been designed that can be dismantled completely into a wooden carrying case 25 × 28 × 13 cm (10 × 11 × 5 in.). The funnel has four sides, and is equipped with legs and two lids—one heated by electricity and the other heated as a water bath by a small spirit lamp. Very lightweight portable funnels can be made of heavy aluminum foil that can be pressed down flat when not in use.

Small arthropods for preservation may be collected by placing 70–80% ethyl alcohol in the jar attached to the bottom of the funnel. Add about 5% by volume of glycerin to the alcohol to help prevent desiccation of specimens in case the alcohol evaporates. The easiest way to collect living arthropods with an extracting funnel is to place moistened crumpled paper toweling instead of preservative in the collecting jar. The arthropods by this means are provided with both moisture and hiding places, which reduce mortality caused by desiccation and predation.

Collect and transport samples for extracting in bags or sacks that can accommodate 14–28 dm³ (1/2–1 cu ft) of material. Polyethylene bags are suitable for extremely wet samples that are going to be held for a few hours only. Otherwise, a cloth or heavy paper sack is better, because air can move freely in the contents of the sack, which at the same time eliminates the problem of condensation or sweating on the inside surfaces of the sack. Cloth bags are best for strength, resistance to moisture, and reuse. Some kinds of samples, particularly humus and soil litter, may be kept in sacks for a week before extraction, with almost no deleterious effect. Some workers routinely predry such samples for several days before placing them in a funnel. Processing of predried samples takes only half as long as do damp samples. Refrigerated soil samples can be held for 2 weeks without adverse effects. Thus, samples of habitats can be collected in the field and brought or sent back to the laboratory, where they can be extracted in Berlese funnels when

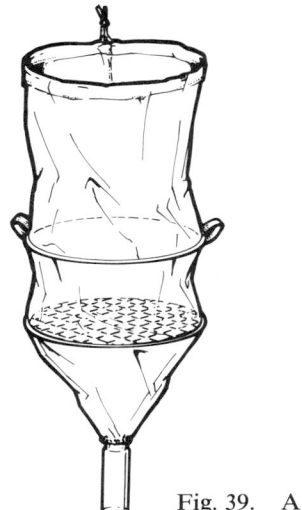

Fig. 39. A collapsible canvas Berlese funnel.

Fig. 40. Berlese funnels on tripodal ring stands with screw-on legs in operation in the field.

it is convenient. Appropriate collection data about the habitat sample should be written in soft pencil on a label, which can be tied to, or placed in, each sack.

One of the variations of the flotation method can be used instead of a Berlese funnel for extracting arthropods from soil, matted ground vegetation, and similar habitats, particularly if there is doubt that the extraction method is going to yield the qualitative results expected of it. The advantages and disadvantages of flotation methods are described in a later section, "Collecting and Extracting Burrowing and Boring Arthropods," p. 56.

Sifter and photoeclector   A sifter (Fig. 41) is one of the most efficient methods of systematically collecting many groups of small, obscure arthropods, particularly Coleoptera, from litter and debris. It is especially useful for separating much of the coarse debris from a sample and concentrating the catch with the finer sifted material for extracting. The sifter is made from two hoops of heavy metal about 30 cm (1 ft) in diam, each provided with a handle. One of the hoops is welded close to the edge of a circular 8 mm (5/16 in.) mesh wire screen of about the same diam. One end of a 1.2–1.4 m (4–4-1/2 ft) heavy canvas cylinder is sewed to the open hoop, the hoop with the screen is sewed into the cylinder about 30 cm (1 ft) lower, and the cylinder is tapered to the end. Tie the tapered end shut with a heavy leather string during use. The handles are positioned so that the lower handle is at about 90° to the right of the upper handle (reverse the position for left-handed collectors). For sampling fine debris, carry a few extra hoops with smaller mesh screens. The most convenient size of screen for most habitats is 5–6 mm (3/16–1/4 in.) mesh. For sampling coarse debris, place the smaller screens in the sifter over the permanent screen. Partly fill the sifter with the material to be sampled, then vigorously shake it to and fro with a slight up-and-down motion to shake the insects through the screen. Remove the coarse material remaining in the sifter. The tied end allows you to accumulate the catch from several samples of a habitat in the sifter. Use the sifter for sampling both wet and dry samples, but be careful to squeeze out excessive water from wet samples such as sphagnum moss, wet leaves, or seaweed, and pull the sample apart in the sifter. Insects may be removed from the sifted residue either in the field or when carried back to the laboratory for extraction. Always store and carry samples in cloth bags rather than plastic bags. For field extraction, use a heavy white plastic sheet about 1.2 × 1.5 m (4 × 5 ft). Spread out this groundsheet on a flat, relatively smooth area and evenly distribute several handfuls of sifted litter over its surface. Pick up insects either with fine, flexible forceps or with an aspirator. If the temperature is low, spread the sheet in the sun to activate the insects, and if the temperature is high, put the sheet in the shade.

A rapid and efficient technique for hand-sorting is to place the sample in a large metal pan and then warm the pan over low heat on a stove or hot plate. As the sample heats up, the insects become more active and come to the surface of the sample, where you can pick them up with flexible forceps or an aspirator. The most efficient method of sorting a sample is by using a Berlese funnel (Fig. 36) or a photoeclector (Fig. 42). The basic construction

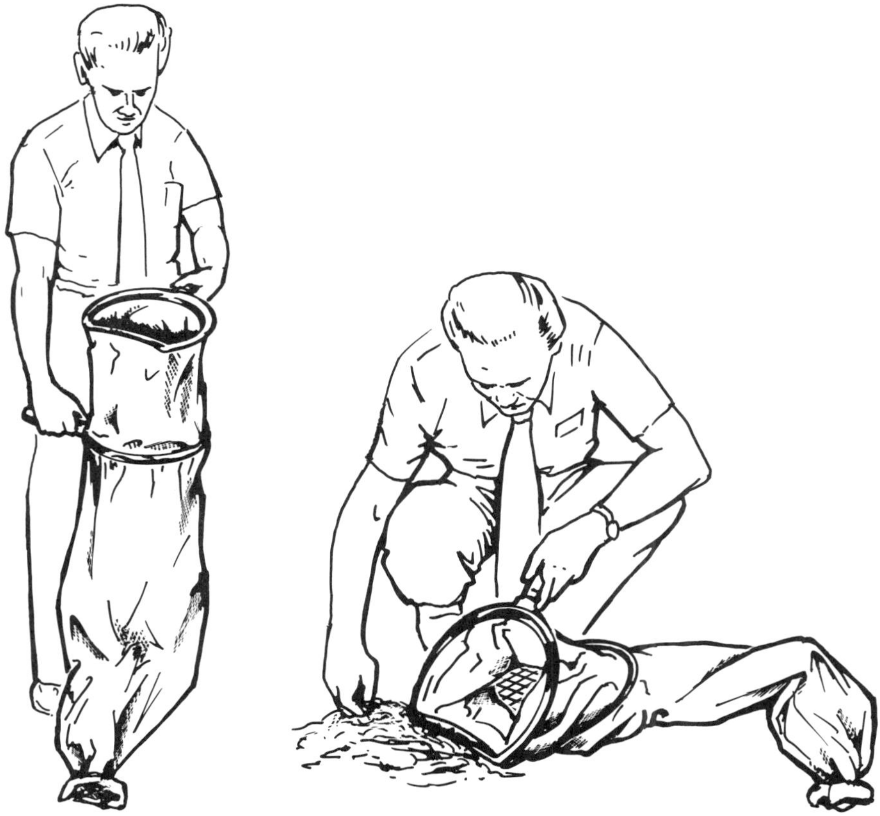

Fig. 41. A sifter operating (*left*) and collecting (*right*) a sample.

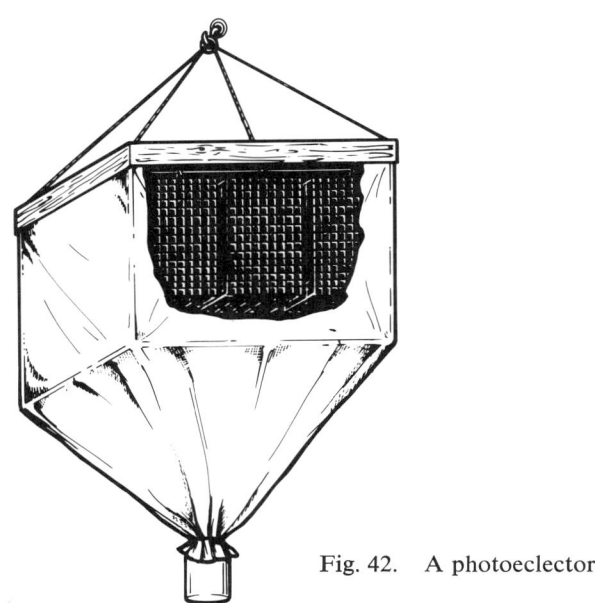

Fig. 42. A photoeclector.

of a photoeclector, as shown in Fig. 42, is a metal frame about 46 cm high, 46 cm wide, and 41 cm deep (18 × 18 × 16 in.) covered with a dark heavy cloth or duck. The top may be a tight-fitting lid or the cloth may extend over the metal frame so it can be tightly tied closed. The lower part of the covering tapers to a ring, which is sewn in to hold a wide-mouth jar. Reinforced wire mesh (mesh of about 5–6 mm, or 3/16–1/4 in.) baskets for holding the sample are suspended by hooks from the top of the frame. The sifted material placed in the wire baskets slowly dries out on the darkened upper part of the apparatus. The insects are attracted to the light and to the higher humidity in the bottom of the apparatus, which is a wide-mouth jar with a moist cloth in it. Check the jar for insects at least once a day and keep the cloth moist. Photoeclectors are slower than Berlese funnels (exposure for at least 2 weeks), but the material can be collected alive and then killed in the usual way, making it easier to mount than specimens killed in a liquid. Also, this apparatus allows the pupae and late-instar larvae that were in the sifted material to mature and be collected as adults.

Collecting and extracting burrowing and boring arthropods

Many of the insects that live in the soil or among the matted roots of plants or that bore or mine in stems, roots, leaves, seeds, or flowers of plants are free-living as adults and may be collected by the methods described previously. Special techniques, however, are needed to collect the early stages of these insects and those species whose adults are not active.

Insects, mites, and other arthropods may be collected from soil or matted vegetation by flotation. Break up a sample of material in a basin of water and stir the water gently, causing the insects to float to the surface, where they may be collected on a sieve or on filter paper. A waterproof bag (Fig. 43) made of canvas, which can be folded for carrying, is especially useful for this purpose. The arthropods separate from the soil easier and faster in a solution of one part magnesium sulfate to three of water or in 1 kg (2-1/2 lb) of granulated sugar to 4.5 litres (1 gal) of water than in water alone. The use of a sugar solution is less harmful to the extracted arthropods than are the other mixtures. More elaborate modifications of the flotation method have been designed: separating the arthropods from plant debris by shaking the "float" in a mixture of benzol and water; causing the nonwettable arthropods in the "float" to adhere to surfaces coated with a grease film such as lanolin; or chemically dispersing the soil.

Flotation methods give more complete and unbiased extractions than do Berlese extractors, because they do not depend on the reaction or mobility of the arthropods to a given stimulus such as desiccation or a repellent. By these methods, eggs, pupae, and other inactive stages can be collected. However, these methods are tedious, the mechanical processes of washing and sieving may damage delicate specimens, and specimens cannot be recovered alive for rearing. Also, arthropods containing a great amount of organic matter do not separate well because not all the debris is disposed of.

Fig. 43. A canvas waterproof bag.

Based on the same flotation principle, you can wash insects from their burrows by saturating the soil near the margins of lakes or streams; you can drive arthropods from the vegetation to the surface of the water by pushing the vegetation under water, for example, by treading on a quaking bog until the area around your feet sinks below the surface; and by stirring up waterlogged mud, you can disturb the insects it contains.

A garden trowel is useful for digging insects from sand or soft earth and a mason's hammer or small geological pick may be used to break up hard soil or pry up rocks. A method of studying populations of soil arthropods in situ is to fast-freeze core soil samples as soon as they are collected. Later, you can thaw the samples in Formalin vapor, submerge them in agar, and section them after they set. By this method, you can kill and fix arthropods in their vital positions and you can observe and record their numbers, habits, and associations.

Most of the insects that bore or mine in plants have to be searched for in order to find them. You need a sharp knife to dig them out and to split twigs and stems; a small axe to split stumps and rotten logs; and a bayonet, small crowbar, or entrenching tool to tear off loose bark or split wood. Many mites can be found in the tunnels of insects that bore under bark or into wood. Do not overlook the usefulness of the Berlese funnel for extracting active arthropods from pieces of loose bark, split wood, rotting logs, lighter organic soils, and matted vegetation. The method is described thoroughly in a preceding section, "Collecting and Extracting Arthropods from Debris," p. 47.

Collecting aquatic insects and mites

Collecting insects in water is somewhat similar to collecting them on land because you use a similar apparatus with a net, but it is modified to cope with the higher resistance of the water.

The net rings described previously (p. 12) may be used for collecting aquatic insects, but special modifications will produce better results. It is important to use a net that offers as low as possible resistance to the water.

If you are capturing only medium-sized or large insects, select a material for the bag that has meshes just large enough to hold these insects. For smaller insects, use a finer mesh and a smaller size of ring; a net with a 20 cm (8 in.) ring is more convenient to use than one with the usual 38 cm (15 in.) ring. Because most aquatic insects do not escape readily from a net after being taken from the water, the bag does not need to be deep; it may be shallower than the diameter of the ring. The bottom of the bag should be truncate or broadly rounded; otherwise a large mass of debris may collect at the point. If the ring is held to the handle by a sliding metal sleeve (p. 00), the last 5 or 6 cm (few inches) of the handle should be sheathed in metal or impregnated with wax to prevent the water from swelling the wood and thereby making movement of the sleeve difficult or impossible. The length of the handle should depend on the way the net is to be used. A handle 1–2 m (several feet) long is convenient for most aquatic collecting. Heavy scrim or the materials used for aerial net bags are suitable for ordinary aquatic nets. Bolting silk may be used for a plankton net (Fig. 44), which has a jar or vial at the bottom where the insects gather.

The ring of an aquatic net may be square with the handle attached to one corner (Fig. 45). This shape is easier to push into masses of aquatic vegetation or between rocks. A more elaborate and more efficient development of this type of net is the apron net (Fig. 46). This shallow net is pointed in front so that it can be pushed through dense growth of aquatic plants, and it has coarse-meshed wire netting over the top to keep out the aquatic plants and to allow the insects to enter. The bag is made of heavy, 3 mm (1/8 in.), rustproof bronze wire mesh, protected in front by a metal sheath. A net with a semicircular ring and the handle attached opposite the straight edge (Fig. 47) is useful for scraping the bottoms of lakes or streams. An improvement on this design is the scraper net (Fig. 48). The net, frame, and handle (Fig. 49) are ruggedly constructed. This net is useful for collecting many of the aquatic groups. All metal parts are machined and the handle is of fiber glass. The net attaches to the frame with snap catches so that nets of various mesh sizes can be replaced easily.

Most water beetles can be collected by using two strong kitchen strainers, one about 18 cm (7 in.) in diam and having about 7 meshes/cm (17 meshes/ in.) and the other one smaller with a finer mesh.

If you stir up the bottom of a stream with a stick, the insects disturbed from the mud and vegetation are carried downstream by the current, where you can catch them in a large net or a screen held at right angles to the current. A screen (Fig. 50) is made of a sheet of fine-meshed copper or galvanized iron netting attached to two stakes. Push the stakes into the bed of the stream so that the screen is upright and at right angles to the current.

One of the most effective ways of collecting immature stages, and at the same time learning something about their habits, is by wading in the water and hand-picking them from stones, waterlogged wood, aquatic vegetation, and other such objects. Moreover, you do not disturb the habitat, which may be important if types of habitat are limited and you need additional material.

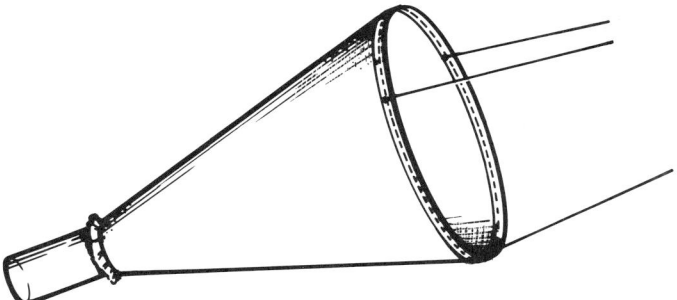

Fig. 44. A cone or plankton net.

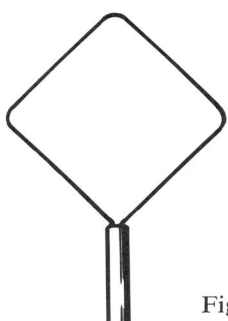

Fig. 45. A square net ring.

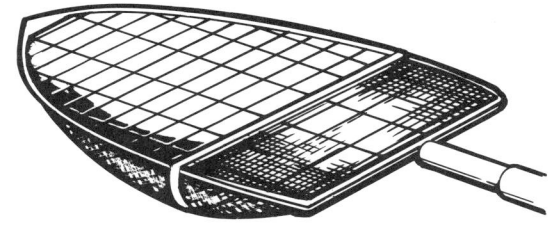

Fig. 46. An apron net.

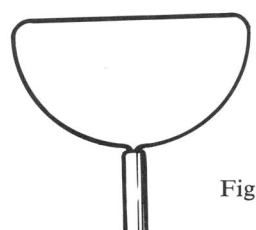

Fig. 47. A semicircular net ring with a straight edge.

Fig. 48. A scraper net.

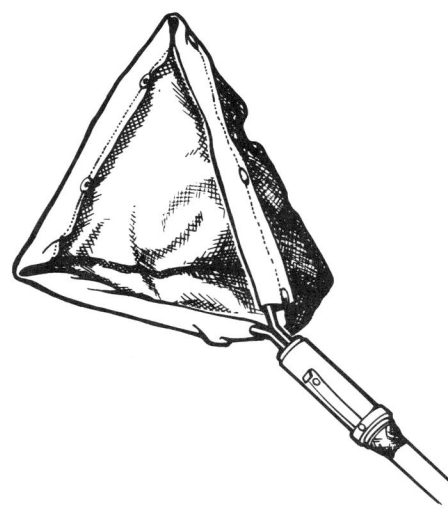

Fig. 49. An aquatic net, frame, and detachable handle.

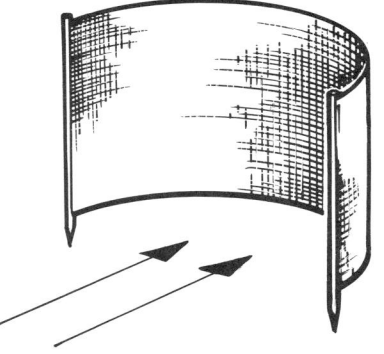

Fig. 50. A screen net attached to two stakes.

A dipper with a panel cut out near the upper edge and replaced with wire gauze (Fig. 51) is useful for collecting insects such as mosquito larvae. The wire gauze allows surplus water to drain off and prevents it and the insects it contains from slopping over the edge. A dipper that can be screwed to a long handle is advantageous. A kitchen strainer is also effective, though larvae are difficult to pick out undamaged and are sometimes hard to see when you remove the strainer from the water. Most small and medium-sized insects can be seen readily in a strainer only when it is held with the open side upward in a pan of water. The flying adults of insects that develop in aquatic habitats may be captured by methods previously described: netting, use of light and other traps, and emergence cages.

The methods described previously for collecting aquatic insects are not entirely effective for aquatic mites, because of their small size and their secretive habits. The methods used for collecting aquatic mites depend largely on whether they occur in open water, submerged vegetation, or on the bottom. Free-swimming mites in ponds or slow streams can be scooped up individually with a small dip net, a nylon stocking, or a tea strainer attached to a long stick. Birge cone nets (Fig. 52) were designed for pulling through weedy areas along shores, where many water mites live. Some kinds of freshwater mites occur in profundal zones and bottoms of deep lakes, and these can be collected in dredge samples, such as are obtained by the use of tall Birge–Ekman dredges (Fig. 53). Various drags, dredges, grabs, and other devices have been designed for sampling the beds and bottoms of water bodies. Strain net and dredge samples by putting them through a 3 mm (1/8 in.) mesh and then examine them in a white tray, where they are easily seen and from which they can be removed with an eyedropper.

Fig. 51. A modified dipper.

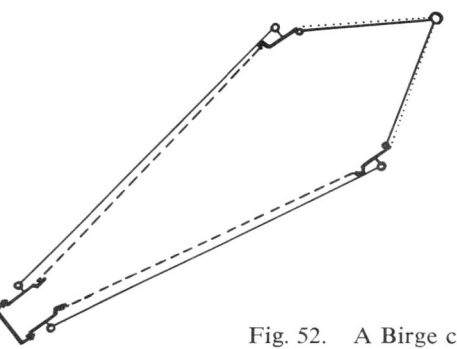

Fig. 52. A Birge cone net.

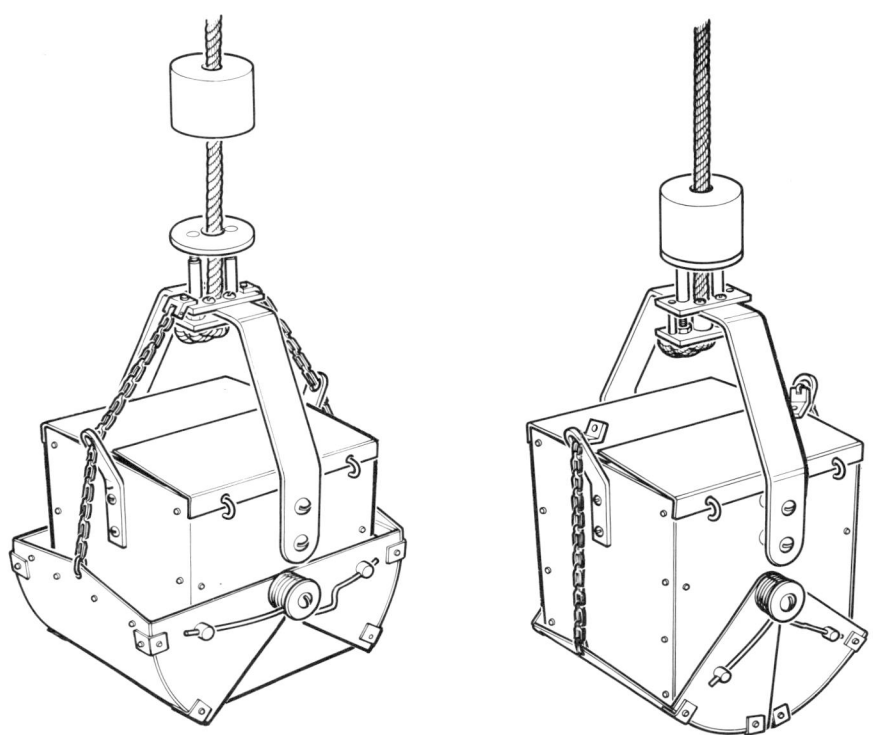

Fig. 53. A Birge–Ekman dredge open (*left*) and closed (*right*).

The larvae of many kinds of water mites are ectoparasites of aquatic insects and mollusks. The larvae and the subsequent saclike nymphochrysalides that remain inside the larval skins are found firmly attached by their mouthparts to the integument of the appendages or the body of aquatic bugs, dragonflies, mayflies, stoneflies, and caddisflies. These larvae are more easily removed intact from their hosts and are more pliable when they are freshly caught or preserved in liquid than when they have been allowed to dry out and become brittle on hosts preserved in a dry state on pins.

Most marine mites are nonswimming and remain tenaciously with their plant or animal hosts, such as algae and coral. Immerse marine samples in a bucket of seawater and add chloroform or a halogenated ether compound as an anesthetic at about 3 kg/m^3 (1/2 oz/gal). Mix or shake the treated water and sample to speed up the reaction of the mites, which relinquish their hold on the substrate. After about half an hour, shake the substrate material vigorously in the water and then remove it from the water. All precipitate in the water can be retrieved by filtering the water through a muslin bag. Then place the precipitate in a container with 90–95% alcohol and label it. You can study and sort it later.

Some unusual marine mites have been collected from samples taken from the bottom of deep ocean basins.

A Riedl dredge with a finer net sewn into the cod end of the 0.5 mm mesh bag has been used to obtain a soft clay bottom sample containing a few halacarid mites, crustaceans, nematodes, and foraminifera.

Aquatic traps Most of the aquatic traps sample a limited area and have been designed for ecological studies or for capturing particular groups of insects. As a result, their use by collectors is limited, although several such traps could be used to supplement general collecting. These traps include the submerged aquatic trap, the surface cone trap, and the simple cone or net trap, which is used to strain the surface of both standing and running water. Grab-and-drag dredges are effective in capturing the immature stages dwelling in or on the bottom. Aquatic traps and sampling devices are referred to briefly in this publication. For further information see Southwood (1966), Edmondson and Winberg (1971), and others listed in the reference section at the end of this publication.

Lentic environment traps A funnel trap (Fig. 54) is most commonly used as a submerged emergence trap. It consists of a funnel with a glass jar at its apex. As the trap is lowered into the water, air is trapped in the jar. Ascending pupae entering the funnel pass up into the jar, and adults emerge into the enclosed air space. Empty the trap daily to prevent the insects from molding, or use a collecting jar containing a solution of alcohol and Formalin to preserve the catch. If you use the latter method, fit the jar opening with a tube (Fig. 54) through which the insects enter and which separates the fresh water from the preserving fluid.

The trap is useful only in water more than 1 m (3 ft) deep; it is not practical deeper than 6 m (20 ft) owing to loss of air in the jar as a result of water pressure. The trap captures most of the aquatic insects living at these depths, retaining both the adults and pupal exuviae. Throughout a season, several of these traps set at various depths will collect samples of the insects living in the deeper parts of a lake. To anchor your traps, suspend them from a line laid out between two fixed buoys. See the list of suggested references for other submerged emergence traps that have been developed on the same principle as the funnel trap.

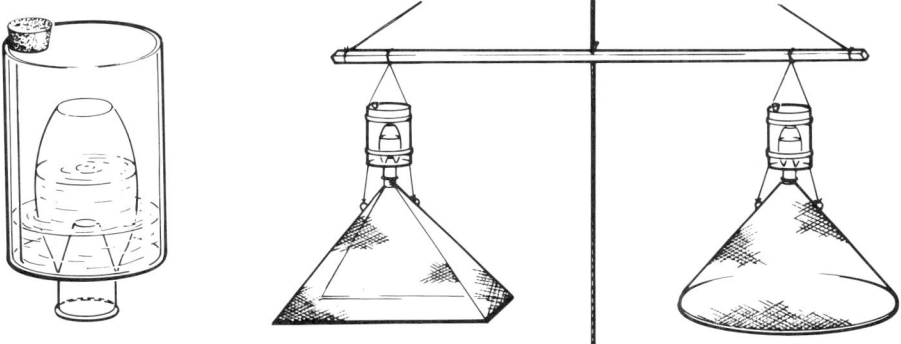

Fig. 54. Funnel traps showing detail of collecting chamber.

Surface emergence traps These traps are designed to capture insects as they emerge from the surface of the water and therefore they can be set over any depth of water. Because they are subjected to wind drag, it is best to use them in the shallow, sheltered margins of a lentic body, which usually have the greatest variety of aquatic insects. A submerged trap is more effective for deeper areas. Most surface traps are difficult to clear, but they can be set and cleared without using a boat.

There are two types commonly used: one rests on the bottom (Fig. 55), and the other is held clear of the bottom either by floats or stakes (Fig. 56). For general sampling, the type held clear of the bottom is best because it captures insects such as mosquitoes that have pelagic immature stages.

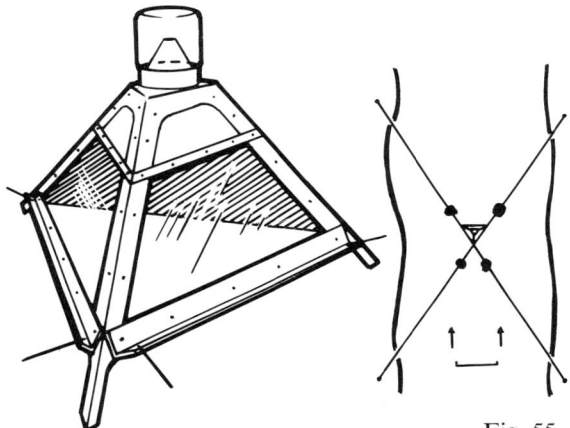

Fig. 55. A surface emergence trap.

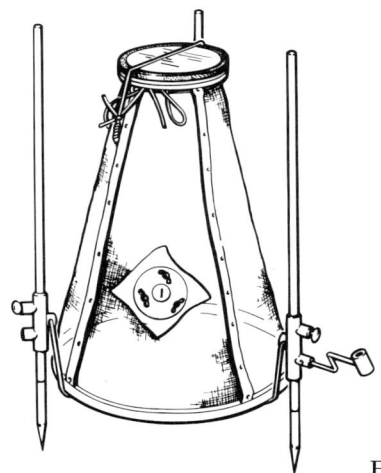

Fig. 56. A surface emergence trap.

Cone or funnel traps have been used successfully; but a box or square trap is more efficient. The catch is collected in a jar at the apex of the trap (Fig. 55) or is removed by an aspirator through a hole (Fig. 56) in the side of the trap. The latter method has been more successful in obtaining insects in northern regions, because many insects do not pass up into the jar. These traps can be held clear of the bottom by floats or stakes.

Tow traps   A single cone or plankton net (Fig. 44) towed through the water so that it is half out of the water captures many aquatic insects, both adults and immatures. It is possibly the best method of sampling aquatic environments. In large bodies of water, the net is towed behind a boat. Smaller bodies of water on margins of lakes may be sampled by standing at the edge and repeatedly tossing the net out into the water and towing it to shore. Preserve the samples in alcohol or select some material for rearing.

A more elaborate type of tow trap for independently sampling the surface and water strata below the surface consists of two coarse cone nets (Fig. 57) mounted one above the other in a rectangular steel frame. The nets are tied in the middle to a horizontal arm close to the water. The two arms of the top of the frame keep the nets a constant distance from the boat. Cover the tip of the nets with a coarse screen and the end with a detachable jar. This procedure concentrates the catch and screens out the coarse debris.

Benthonic samplers   These are not exactly traps, but they were designed to capture insects living in or on the bottom of lentic bodies. The most commonly used types are the Birge–Ekman (Fig. 53) (for soft bottoms) and the Peterson (Fig. 58) (for hard bottoms) dredges or grabs. The movable jaws on these grabs close when they reach the bottom and hold and pick up the enclosed sample of bottom sediments. Core samplers of various types can be used, but the samples they collect are too small for general collecting.

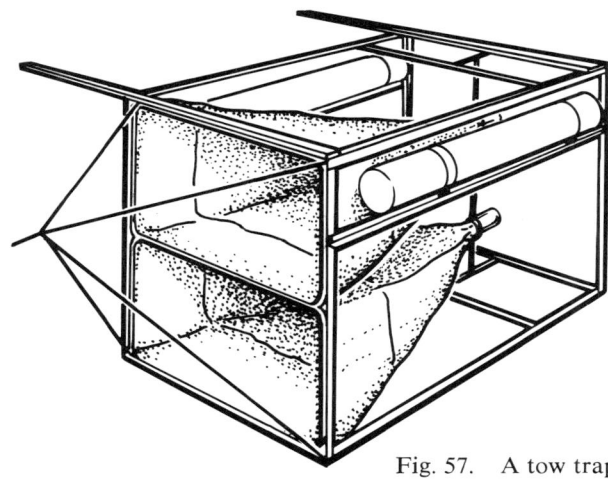

Fig. 57.   A tow trap with two coarse cone nets.

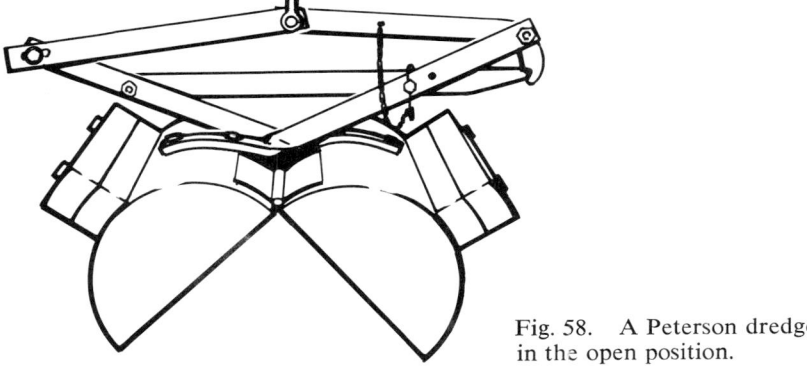

Fig. 58. A Peterson dredge in the open position.

Nets supported on metal frames dragged along the bottom effectively collect organisms from the bottom of lakes or rivers. The leading bottom edge of the frame is constructed so that it cuts down into the substrate as the net is pulled along. Some of these dredges have a sifting arrangement at the end of the net for separating the organisms from the substrate. Smaller types of drag dredges can be operated from the shore by repeatedly tossing the dredge out into the water and dragging it back along the bottom. Another method used for sampling the bottom from the shore is a metal sieve with a long handle that is used like a rake. These drag dredges only function on soft bottoms. Tines may be welded across the mouth of the drag to disturb the substrate and to prevent large stones and debris from entering the bag. It is almost impossible to sample rocky bottoms that are any deeper than 1 m (3 ft) other than by diving. To sample shallower rocky bottoms, enclose an area in a round or square sampler extending from the bottom to above the water surface. This enclosure prevents insects from swimming away. If the bottom is fairly loose, push a metal sheet under the bottom of the sampler and lift out the sampler and its contents, or pick up the rocks and gravel by hand and strain the enclosed water with a small hand net.

Lotic environments The use of cone stream traps is probably the best way for a general collector to sample a stream. Suspend a cone or plankton net (Fig. 44) in a stream in such a way that it is half out of the water. It catches all the insects, both adults and immatures, floating down the stream. The tip may be fitted with a screen and bucket to separate the specimens from the debris.

Emergence cages, described in an earlier section of this publication, may be used in a slow-flowing stream. Funnel traps are more efficient than boxlike traps. All these traps are hard to empty because they tend to gather debris at their upstream edge, which may damage the specimens. A triangular stream trap overcomes the previous disadvantage. It consists of a metal frame covered with metal gauze and has the usual glass jar at the apex with an inverted funnel. One angle of the triangle facing upstream prevents the accumulation of debris, but unless the insects rise into the glass jar, they are hard to get out. This trap can be set up on legs so that the water flows under it.

Collecting ectoparasites of vertebrates

Birds, mammals, and reptiles are attacked by fleas, lice, mites, and ticks, as well as by certain parasitic flies, bugs, and beetles. Not only terrestrial but also aquatic vertebrate animals may be hosts. To collect parasites you have to capture and usually kill their hosts, either by shooting or trapping them, though useful material may sometimes be obtained from animals found dead on roads. Rodent snap traps of the Museum Special type usually do not crush the animal's head, and this facilitates later examination of sinuses and nasal passages for respiratory mite parasites. Similarly, the use of a bird shot (sh t of small size for shooting birds) for bird collecting reduces damage to the host and makes the examination for parasites easier. Further details on collecting mammals and birds and preparing study skins, if these are required, are given by Anderson (1948). If you are interested in collecting animals and birds, be sure to find out and to comply with the provincial and federal regulations regarding shooting or trapping of the species you want. Even when you obtain a permit, always use the utmost discretion and do not unnecessarily molest or kill animals or birds.

Put the dead bird or mammal in a paper, polyethylene, or cloth bag and close it to prevent the escape of ectoparasites. Do not put more than one species of hosts in the same bag. Place a small wad of absorbent cotton soaked in chloroform in the bag and after a minute or two the fumes will anesthetize the parasites. Search the animal carefully and comb it thoroughly with forceps. A vigorous combing usually loosens fleas, larger mites, and other active forms. Ticks usually do not detach, but must be searched for and picked off. Some of the smaller lice and mites are difficult to find by searching and you may have to dissolve the fur (p. 129). Mites on dead rodents and birds can also be removed by washing and shaking the host in a mixture of water and detergent. Decant the detached ectoparasites into another container from which they can be collected with the use of a brush, eye dropper, or forceps, or they can be filtered out with the aid of a Buchner funnel and a sink vacuum.

A successful method for removing lice and other ectoparasites from a host is by thoroughly dusting the host with the finely-powdered silica Dri-die. This dust breaks the oil film of the ectoparasite's integument and causes rapid dehydration and irritation. Most ectoparasites drop from the host within a few minutes after the dust has been applied.

A direct examination of vertebrate hosts is necessary for collecting some mite parasites, and a stereomicroscope greatly facilitates this search. First, carefully examine the external surfaces of the host. Tiny mites that are firmly attached to fur and feathers may be easily overlooked. The body areas covered by the wings and the flight feathers on the wings are the favorite spots for mites on birds. Collecting mites that are living within feather quills can only be done by dissection. Skin mites are hard to locate, particularly if the infestation is subcutaneous, and it is necessary to spread back the fur or feathers of the host in such a way that the skin of different body areas is clearly exposed. Next, examine externally exposed organs, particularly the ears, eyes, nose, and genital and anal openings. Finally, examine certain internal organs such as the lungs, sinuses, and respiratory passages.

Collecting ectoparasites from living animals can be done in several ways. Rodents can be caught with baited live traps, bats and birds with mist nets, and larger mammals temporarily subdued by anesthetic projectiles. Vigorous brushing and combing is one of the simplest and most effective methods of collecting from mammals. Smaller birds and mammals may be kept for a few days in a screened cage suspended over a pan of water, and many of the ectoparasites will detach from the hosts after feeding and drop into the water, where they may be collected. Captured living reptiles can be isolated in the same way for collecting ectoparasites. A modification of this method is to cage the hosts above Berlese funnels, without using a desiccant or repellent. Many of the ectoparasites automatically drop down and collect in the jar below; the autosegregating grid retains most of the feces and urine, and should be cleaned often. Internal parasites and obligate ectoparasites that do not drop off their hosts after feeding are not effectively collected by these caging methods.

For collecting ectoparasites from living birds, the Fair Isle apparatus may be used. By this method, the bird is enclosed in a cylinder but its head and neck protrude through a hole in the upper lid. After chloroform vapor has been pumped into the cylinder, the bird is stimulated to flutter so that the anesthetized parasites drop onto the suitably colored surface of the base of the apparatus. A suitable sleeve around the neck of the bird prevents the chloroform from reaching its nostrils.

The nests of birds and mammals are usually productive sources of fleas, mites, and ticks, and they may also be inhabited by various arthropods, not ordinarily collected in other habitats. All these insects may be collected by the methods previously described, especially by sorting through the nest material over a white pan (p. 35), or by using a Berlese funnel (Fig. 36). The litter along mammal runways and around bird perches may also yield ectoparasites when it has been extracted by a Berlese funnel.

Unattached ticks can be collected from vegetation by "flagging" or pulling a "tick drag" through an infested area. The drag (Fig. 59) consists of 1 m^2 (about 1 sq yd) of light-colored flannelette or wool cloth, tied by tapes like a flag to a stick about 1.5 m (5 ft) long. As you walk through the area holding the drag in front of you, place it on likely pieces of vegetation first on one side and then on the other. As the drag moves slowly over the grass and bushes the ticks voluntarily leave their perches and drop on to the drag, from which they may be picked off and placed in dry shell vials with corks (if you want to keep them alive) or in alcohol vials. Living ticks are awkward to handle and you need some experience before you can place one in a vial without having the others crawl out. If you jiggle the vial the ticks will bunch together at the bottom of it, and they will often "ball up" with each other and with bits of lint from the drag. Larval ticks of some species may be collected when at certain times they form clusters of hundreds (the progeny of one female) on a single blade of grass. Dragging for ticks is impractical on cold, wet, or windy days.

Newly hatched chiggers tend to move onto any new object placed in their environment. Black or white disks, dishes, trays, or oilcloth squares left on

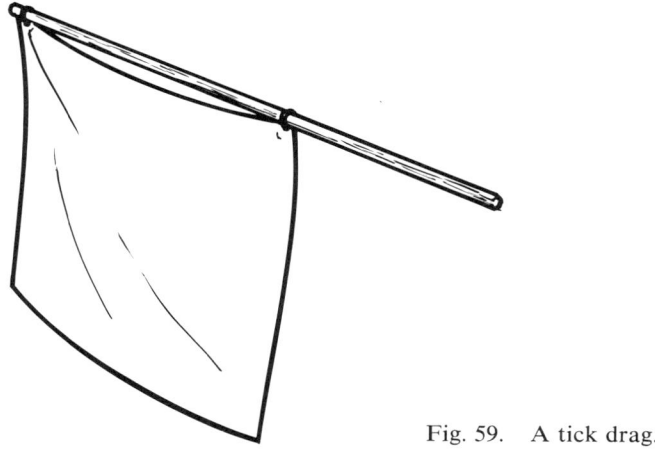

Fig. 59. A tick drag.

the ground for several minutes will attract specimens that can be removed and placed in vials. More elaborate and specialized devices for collecting chiggers, based on positive phototropic response, are also useful.

Collecting mites associated with invertebrates

A wide variety of mites can be found on many different kinds of terrestrial and aquatic insects, arachnids, and myriapods, and less commonly on crabs, mollusks, sea urchins, and sponges. Some are true parasites of their hosts, whereas others are commensals or simply phoretic associates. The mites are most easily found after their hosts have been collected and killed. They are often peculiarly localized on their insect hosts, their favorite spots including on the pronotum; on the abdominal sternites; between the coxae; under the wing covers; on the abdominal tergite conjunctiva; appressed in depressions on the head, thorax, and abdomen; in spiracular cavities and their main tracheal trunks; and attached to the legs and wing bases. Insects collected in alcohol remain pliable for searching for mites under their wing covers, in their spiracular openings, and in other recesses where some prying or spreading of parts with forceps may be necessary. However, because some kinds of external mites wash off their host's body, carefully examine the alcohol for mites and never discard or replace it, especially if it is the original liquid in which the collection was made. Also, collect only one kind of host in each alcohol vial, in order to avoid mixing data on host association. Search for mites on insects collected dry in killing bottles before the host and mites dry out and become brittle. Once they have become dry, the mites cannot be placed directly in alcohol for preservation.

Rewarding collecting for mites may be made on insects in museum collections, providing you take great care in handling the dry, fragile hosts. If necessary, the insects may be relaxed in humidity chambers to minimize breakage. Dry mites may be mounted directly in Hoyer's medium or wetted with a droplet of glycerin and then removed with a pin or fine brush and placed in alcohol.

Certain parasitic mites, such as gill parasites of pagurid crabs, ectoparasites of sponges and mollusks, and internal parasites of deep-sea urchins, usually will be discovered only by specialists with a thorough knowledge of the hosts.

In addition to collecting directly on invertebrate hosts, the mite associates may be collected from the habitat occupied by their hosts. This is particularly rewarding in the case of social and subsocial insects who build nests or make tunnels in which a diversified biota, including mites, coexists. The methods previously described, of sorting through the habitat material on a white tray and of extracting with a Berlese funnel, are effective.

Rearing

Among the purposes of rearing living insects and other arthropods are to obtain specimens in perfect condition for taxonomic study, specimens of early stages associated with adults so that the former may be identified and described, representatives of species that are never or rarely captured by usual collecting methods, and parasites with host data. Another reason for rearing insects is to observe them and to do experimental studies on their bionomics. Rearing methods and equipment form a major aspect of entomological work. Peterson (1949), in *A Manual of Entomological Equipment and Methods,* illustrated or referred to over 400 kinds of outdoor or indoor rearing, breeding, hibernating, feeding, oviposition, emergence, and observation cages or other containers. Rearing equipment for insects and arachnids is a subject that is too extensive to be discussed adequately in this publication. For insect-rearing equipment see Peterson (1959) and for culturing techniques for mites refer to Evans et al. (1961). However, the chief principles of successful rearing are given in the following pages.

Living insects or arachnids are usually carried home from the field in metal, glass, or cardboard containers that are small enough to be carried conveniently. These are often unsuitable as rearing containers, because conditions fatal to the specimens can develop rapidly within the containers if suitable precautions are not taken. Deaths occur usually as a result of excessive moisture or dryness and lack of food. Moisture from the insects' bodies and from their food plant condenses on the inside of metal or glass containers and may kill the insects. Avoid this occurrence by keeping the container in a cool place, not overcrowding the specimens, and putting a piece of dry blotting paper in the container. Small holes or panels of wire screening or perforated zinc in the walls of the container or a muslin or absorbent cotton covering instead of a solid stopper permit some air circulation and help to prevent saturation of the air. A finer cloth screening is needed to prevent smaller arthropods, particularly mites, from escaping. The danger of excess moisture can be avoided by using a cardboard or wooden container, but in such containers dryness may develop rapidly to an undesirable degree. Moreover, some insects are able to eat their way through cardboard, or wood. Too rapid drying may be prevented by enclosing the food plant or other foliage with the insects. Never expose a container to the sun or carry it where the heat of the sun can reach it, because heat is the chief cause of rapid drying or condensation, and usually is the direct cause of death. Supply the insects or arachnids with enough food to last them until more supplies can be obtained. Even if you do not intend to rear the specimens, it is wise to collect some of the plant or other material with which they are associated. This plant material delays the specimens from drying out. The identification of these plants may be confirmed later by more careful examination than is possible in the field. Also, the plants supply oxygen for aquatic species and provide something to cling to when the insects are being carried. For soft-bodied insects, spiders, and large mites, put the plant or other substrate into the container first, to prevent crushing the specimens accidentally. Do not put more than one species in the same container, because they might attack one another. It is sometimes unwise to put more than one specimen in a container. Transfer the specimens without delay from their temporary containers to suitable rearing cages. Select the kind of rearing cage best suited to the habits of the arthropod. Insects and arachnids require certain conditions of food and environment, and species vary in their tolerances for unfavorable conditions. Successful rearing involves reproducing the favorable, and if possible the optimal, conditions and reducing those that are unfavorable. The difficulty in reproducing optimal, or even acceptable, conditions varies with the species. In general, arthropods that have a wide range of environmental tolerance—those that live exposed to weather changes and that inhabit a variety of habitats—are easiest to deal with, because their range of environmental tolerance includes the unnatural conditions of a rearing container. Some insects, for example many aquatic, fossorial, and wood-boring species that live in relatively stable environments that can easily be reproduced or kept stable in a container, are also easy to rear. However, some insects and arachnids that have more restricted ecological requirements or whose natural requirements are difficult to reproduce or to keep stable in the laboratory are less easy, and often extremely difficult, to rear successfully. In rearing these or any arthropods, a lot of experimenting may be needed to achieve the best results.

The atmosphere inside a container always differs from that outside of it: the temperature and humidity tend to be higher and the air movement and amount of light lower. These factors vary greatly, depending on the location and size of the cage and the material from which it is constructed. The atmosphere inside a cage of netting or wire screening more closely approaches that of the insects' natural habitats outdoors than that in a glass container in a laboratory. However, the specimens are easier to observe and they escape less readily from a glass container.

A point sometimes overlooked is the necessity to ensure that environmental conditions during the hibernating stage, as well as during active stages, resemble those in nature. Though an unbroken sequence of generations of many species can be reared in the laboratory, some species require a period of cold in order to complete their life cycles. It has been found that some hibernating insects and arachnids often survive better when they are exposed to the weather in an outdoor cage than when they are kept indoors. When most hibernating pupae are kept in cold storage at 0–4.5°C for about 3 months, then are removed and placed in an incubator or rearing room, they usually emerge as adults (hosts or parasites).

The specimens must be supplied with plenty of fresh food. Otherwise, their growth may be retarded, their behavior may become abnormal, they may become more susceptible to disease, or they may die. Fresh food should be supplied before the food from the previous feeding starts to wither or dry up. It is always better to cage phytophagous insects and mites on a growing plant than to give them cut foliage.

Besides reproducing the natural environment as closely as possible and supplying plenty of fresh food, other requirements of the insects or arachnids should not be overlooked. For example, you must provide soil for species that pupate, oviposit, or hibernate in the ground; suitable materials for insects that construct larval cases or cocoons; and twigs for those species that oviposit or pupate on such materials, or that need something to cling to while they are drying and hardening after emergence, or for spiders that need a backing on which to build a web.

Parasites and predators are among the unfavorable factors in the natural environment of an arthropod. Their attacks are reduced automatically when the host specimens are enclosed in a cage or other container. Also, attacks by predators, which may be an important control factor in the field, are excluded from a cage, unless the predators are small, such as mites, or are introduced accidentally with the food plant. Attacks by insect and mite parasites are also reduced, but insects may die from this cause if they were attacked before their capture. The likelihood of attacks by bacterial, viral, fungal, and protozoan diseases, however, may be increased in rearing containers because of their unnatural environmental condition. The closer the environment in a container is to the optimal conditions for the species being reared, the lower the death rate will be from disease or from more direct causes.

Because the life histories and behavior of arthropods are influenced strongly by their environment, and the environment within a rearing container always differs in some respects from the natural environment, misleading conclusions may sometimes be arrived at if data obtained by observing specimens in the laboratory are applied to the arthropods in the field. Laboratory observations should supplement, not replace, field observations.

Do not kill adult insects and arachnids that have been reared immediately after emergence. Keep them alive until their bodies, wings, and other appendages have expanded and hardened and their full colors have developed. This development may take from a few minutes to a week or more, depending on the species.

Preserve the remains of the hosts from which parasitic insects have emerged or the characteristic remains of the insects and arachnids themselves, such as eggshells, larval cases, cast-off larval or nymphal skins, cocoons, empty pupal cases, mined leaves and stems, or galls. Such remains are sometimes needed for identifying the insects or arachnids and they often provide valuable taxonomic or biological information. They can usually be preserved in microvials or gelatin capsules on the same pins or in the same vial as the specimens. Preserve any parasites that emerge from insects and arachnids while they are being reared. As soon as you notice the parasite cocoon, put it in a separate vial, with all the data necessary to associate the parasite with the host. The wide gaps in our knowledge of the taxonomy and morphology of the early instars of insects, mites, and spiders will be reduced gradually as the preserved examples of the early instars associated with adults that can be identified become available. Consequently, when a species is being reared, it is extremely important to preserve examples of each stage of its development.

Predatory arthropods such as spiders can be reared in paper or plastic cups fitted with a screened lid. Use a stoppered side opening to introduce living prey. Provide water by stacking a cup inside another cup of the same size, the lower one containing water, and project a dental wick from the lower cup containing the water through a hole into the upper cup. Because some spiders are long-day maturing, you may have to provide extra hours of light daily during autumn and winter. Most spiders are cannibalistic and predatory and should be reared individually.

Killing agents and killing bottles

A killing agent is used to kill insects as fast as possible, without affecting their colors or hardening them unduly. None of the commonly used killing agents fulfills all these requirements.

The most widely used killing agent is cyanide (calcium cyanide or cyanamide, sodium cyanide, or, preferably, potassium cyanide). Because cyanide is deadly poisonous, many collectors are reluctant to use it. The sale of cyanide is restricted and it is not readily available. If you use cyanide, be extremely careful. It must be used only in a specially prepared killing bottle (of polypropylene or cellulose) that must never be left where there is a possibility of anyone interfering with it. Burn old bottles (used for one season) in an incinerator. Cyanide has certain disadvantages as a killing agent for some insects: it leaves specimens brittle and hard, making them difficult to dissect, and it turns yellow to red, pink, or orange, and some greens to a yellowish color when specimens are left too long in its fumes. Such color changes occur quickly at high temperatures.

A killing bottle (containing cyanide or another killing agent) becomes gradually less effective with constant use. Do not test its strength by smelling it; the length of time insects take to die in it is a good indicator. The bottle should have a tight-fitting stopper or lid, because the fumes are poisonous or dangerous to health.

The handiest and safest cyanide bottle is a tube of unbreakable transparent plastic. The tube may be about 2–2.5 cm (3/4–1 in.) wide and 10–15 cm (4–6 in.) long. Large-mouth jars are best for larger insects, such as large moths or butterflies, and for immobilizing the catch concentrated in the net tip (p. 15).

To prepare a cyanide bottle or tube (Figs. 60–62), place a layer of granulated cyanide about 13 mm (1/2 in.) deep in the bottom of a jar or tube. Potassium cyanide is the least likely to deliquesce. Next, pour sawdust or dry plaster of Paris over the cyanide into the tube until it forms a layer about 2–2.5 cm (3/4–1 in.) deep when tamped down, and add three or four drops of water. Over this put either a layer of cream of plaster of Paris about 13 mm (1/2 in.) deep or for small tubes only, a tight plug of absorbent cotton, which may be covered with a plug of tissue or a tight-fitting circle of blotting paper or porous cardboard. If you choose the former, leave the bottle open out-of-doors for 12–24 hr to allow enough time for the plaster of Paris to dry. If you use absorbent cotton, pack it in tightly and place over it a looser plug of tissue, which can be replaced as it becomes soiled and can be used for cleaning the inside of the tube.

Liquids whose vapors are used to kill insects include ethyl acetate (acetic ether), tetrachloroethane, carbon tetrachloride, ether, chloroform, benzene, ammonia, and ethylene dichloride. They are safer to use and more easily renewed than cyanide, though some precautions must be taken. The vapors of carbon tetrachloride and other fat solvents are harmful to humans, and an excess of them can eventually cause an incurable disorder of the liver. Benzene, ether, and ethyl acetate are inflammable. Carbon tetrachloride and chloroform can give rise to phosgene, a deadly gas, in the presence of a naked flame. Used as killing agents for insects, these chemicals are adequate, but some are inferior to cyanide. Though some of them stupefy the insects quickly, they kill more slowly than does cyanide and the specimens must be left in the killing bottle for at least half an hour. Excess of a liquid killing agent may wet the specimens and permanently damage any that have fragile scales, hairs, or wings. Benzene and ammonia may change the color of specimens, particularly green and yellow ones. Carbon tetrachloride and chloroform leave the specimens hard and brittle, and therefore difficult to mount. However, the former is useful as a killing agent in an enclosed space, such as the container of a light trap, because its vapor is heavier than air and it does not dissipate readily, but tetrachloroethane is better for enclosed areas, particularly because it does not stiffen specimens unduly and does not alter the color of green specimens.

Ethyl acetate is an effective killing agent for Coleoptera, Hymenoptera, and several other groups of insects. It is not recommended for Lepidoptera or other fragile winged specimens. It has the advantage that it cleans grease and dirt from specimens, but does not harden them. Insects may be killed and left indefinitely in tightly closed vials containing a few drops of ethyl acetate. Beetles may be kept in vials partly filled with sawdust as long as the sawdust has been moistened with a few drops of ethyl acetate.

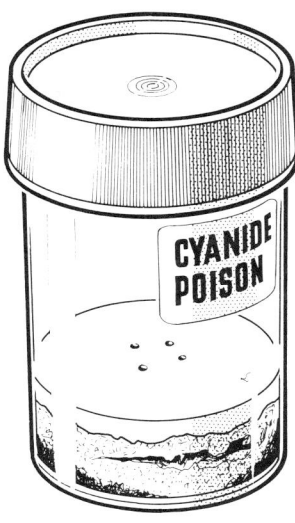

Fig. 60. A cyanide killing jar labeled poison.

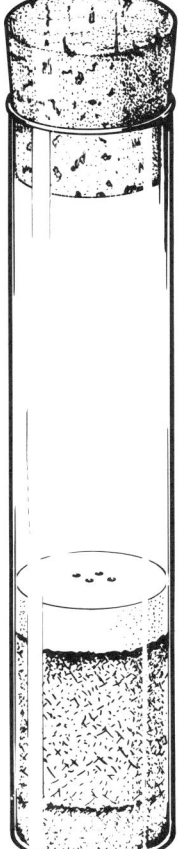

Fig. 61.
A killing tube.

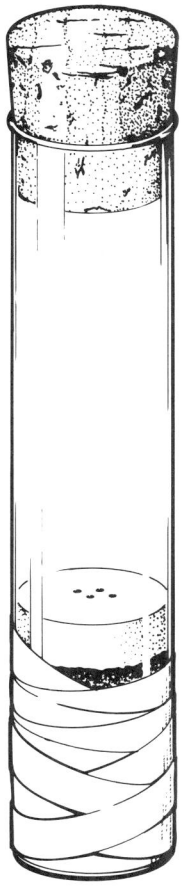

Fig. 62.
A killing tube taped for additional strength.

Any glass jar or bottle with a tight-fitting stopper is suitable for use as an ethyl acetate killing bottle. For field use, a 10.6 mL (3 dram) or larger vial is useful. A supply of vials for 1 day or several weeks of collecting can be easily made by filling each vial about one-quarter full of sawdust. Use large-grained sawdust that has been sifted to remove the fine particles. Hardwood sawdust is best, because ethyl acetate dissolves the pitch and gums found in softwoods, which leaves a residue on the specimens. Place a few drops of ethyl acetate on the sawdust shortly before you intend to use the vials. Specimens placed in the vial burrow into the sawdust and are killed quickly. At the end of each day of collecting, place a label in the vial, press a wad of absorbent cotton down on the sawdust, moisten the cotton with a few more drops of ethyl acetate, and tightly stopper the vial with a neoprene rubber stopper (Fig. 63). The specimens can be left in the vials until you have time for mounting them. If the vials are kept tightly closed, the beetles will still be relaxed for mounting up to a year after they are collected. If the ethyl acetate evaporates, add a few drops to the vial several days before you plan to do the mounting.

Ethylene dichloride (Dichlorethane) is satisfactory as a liquid killing agent for most insects. However, Orthoptera specimens turn pink if they are left too long in a killing bottle containing it.

A killing bottle using ethylene dichloride is easily made. Place a block of sponge rubber in the bottom of a glass tube and allow it to absorb about 10 drops of ethylene dichloride. Press a plug of tissue paper down over the rubber, leaving an air space of about 6–13 mm (1/4–1/2 in.), and add a few thin strips of tissue, loosely crumpled, to restrict insect movement. The tube is then ready for operation. The upper side of the tissue plug should be flat and smooth. This can be achieved by wrapping a single tissue around a core of absorbent cotton. Excess ethylene dichloride causes condensation inside the tube; this should be released by opening the tube for a few seconds before using it. Specimens become brittle if left too long in the fumes. During collecting, remove specimens at frequent intervals and store them loosely in an airtight container with a fresh leaf. A properly charged tube should last several days and it can be recharged as often as necessary.

As mentioned previously in the description on light traps (p. 15), tetra-chloroethane is the safest and most efficient killing agent. Also, it may be used in a killing tube or bottle made in the same way as those used for ethylene dichloride, except that a plug of absorbent cotton should be used instead of the block of foam rubber.

A glass container is best when you are using liquid killing agents because many of these chemicals are plastic solvents. The decision to use a jar or a tube depends on the size of the insects you are collecting: for large insects such as large moths, you need a wide-mouth jar; for smaller insects such as flies, a tube 2 cm (3/4 in.) in diam is better because it can be easily closed with your thumb as specimens are inserted, or it may conveniently fit into the mouth of an aspirator. To make a simple killing bottle or jar, place a layer of an absorbent material in the bottom of a bottle and pour a few drops of liquid on it when the bottle is going to be used. It is advisable to put a few layers of blotting paper over this absorbent layer, to prevent the insects from coming into contact with the liquid. The absorbent layer may be a layer of plaster of Paris 1–2 cm (1/2–3/4 in.) deep and dried thoroughly before use (Fig. 64) or a layer of felt about 6 mm (1/4 in.) deep and covered with a layer of absorbent cotton. For a temporary killing bottle, a layer of sawdust covered with a tight plug of absorbent cotton is satisfactory. Another type is a bottle with a thick layer of absorbent cotton covered with a layer of plaster of Paris, through which passes a glass tube that is open at both ends. When you are going to use the killing bottle, drop 2–3 cm^3 of the killing agent into the tube and plug it with absorbent cotton to prevent the insects from entering the tube.

Fig. 63. A vial containing a specimen of Coleoptera in sawdust and ethyl acetate.

Fig. 64. Killing jars for use with a liquid killing agent.

The absorbent material may be placed in the stopper instead of at the bottom of the bottle. Bore a hole through the middle of a large cork stopper and into it fit a short tube that opens into the bottle. This tube (Fig. 65) contains a layer of plaster of Paris, absorbent cotton, or felt, as described previously for a liquid killing agent. Such a stopper may be used in conjunction with a cyanide bottle to quickly knock down the captured insects, or it may be used to replace the stopper of a container of living insects, for example, the container of an aspirator.

Be careful not to let the insects get damaged or dirty in the killing bottle. Do not overcrowd the specimens; it is better to have a number of small bottles with a few specimens in each than one large bottle packed with a disintegrating and interlocked mass of insects. Tough and fragile, or large and small, do not put these insects in the same bottle, because they will damage one another. Insects (such as grasshoppers and some beetles) that expel juices or excretions may ruin other specimens if they are placed in the same bottle. Specimens of Lepidoptera must not be mixed with other insects, because the latter then become covered with scales and hairs. For the same reason, other insects should not be put in a bottle that had previously contained Lepidoptera before thoroughly cleaning the bottle. It is extremely important to always use clean, dry killing bottles. Clean your killing bottles regularly. A tube-type bottle can be cleaned easily by pushing a slightly dampened plug of absorbent cotton up and down inside it with forceps, and a larger bottle can be cleaned with a cloth.

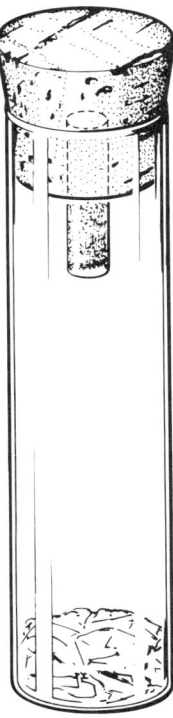

Fig. 65. A killing tube with a stopper holding a vial containing the killing agent.

Nonhairy and nonscaled insects (such as many beetles and larvae) can be killed by immersing them in boiling water. Place them in a beaker of water, bring it almost to a boil, and then let it cool gradually. Arthropods that are kept in liquid preservative do not need to be killed first; they may be dropped into the preservative alive. However, larvae should be killed before they are placed in liquid preservative to prevent distortion (*see* the section "Preservation in Liquids," p. 98).

Clearly print POISON on the label of all killing bottles and containers, especially those containing cyanide. Be sure to write the name of the killing agent on the label. It is wise to put adhesive tape on both the upper and the lower parts of the bottle, if the bottle is large, to minimize the danger of breakage.

Equipment and methods for preserving and mounting

Incorrect or careless mounting or preserving reduces the scientific value of specimens. The difference between good and bad mounting or preserving is shown when one specimen can be identified readily and another one can be identified only with difficulty or not at all, or, specifically, a specimen of scientific value and one of no value. The standard methods of mounting and preserving various kinds of insects that have been developed from the combined experiences of entomologists the world over are so widely known that it is difficult to attribute the high proportion of badly mounted or incorrectly preserved specimens in many collections to anything but carelessness. The chief methods are described in this publication, and their application to specific insect groups, as well as some specialized methods, is described in the section "Applying the Methods," p. 124.

The method of preservation should be determined by the structure of the insect. Insects may be separated into two groups: soft-bodied forms, whose shapes and structures are altered rapidly after death by decomposition or drying; and hard-bodied forms, whose shapes and external structures are not altered significantly by the decomposition or drying of the soft internal organs. Usually the former are preserved in liquid, whereas the latter are mounted dry without the use of agents to prevent decomposition.

Relaxing

Hard-bodied insects that are to be mounted dry must be in a relaxed condition while being mounted. That is, their bodies and appendages must be sufficiently movable and flexible to be arranged so that the structures necessary for identification are visible or are shown to the best advantage. If at all possible, mount the insects immediately after they have been killed, before they have begun to dry and stiffen. They are easiest to mount satisfactorily at this time, are less liable to accidental damage or to harm from pests or molds, and maintain their colors better. Large specimens may take several days to stiffen; small specimens dry rapidly, but by putting them in an ethyl acetate killing bottle they may be kept for a long time in a condition suitable for mounting; small specimens will keep for a day or two in a salve tin or other tight-lidded container with a few green leaves or some blotting paper and a few drops of water. However, some decomposition of internal structures may take place under the latter treatment and the genitalia may become distorted.

Specimens that have dried out must be softened, or relaxed, in a relaxing box. The box is also used to keep fresh specimens in a workable condition until they can be mounted. Keep the humidity high inside the box. A relaxing box may be made from any fairly large metal or glass container. Those sold commercially are usually made of zinc. A large tobacco or cookie tin is satisfactory; although it rusts through after a while, it is easily replaced. A glass battery jar, a small aquarium, or a large earthenware jar with a sheet of glass over the top is also satisfactory. Use a container that is tall in proportion to the diameter of its opening; otherwise it is difficult to keep the humidity high inside when the specimens are being removed. Place a layer of porous material that is capable of absorbing a great deal of water in the bottom of the container (Fig. 66). This layer should be about 4 cm (1-1/2 in.) deep and may be of sand, absorbent cotton, soft paper, balsa wood, plaster of Paris, fiberboard, or synthetic sponge. Saturate this layer with water. To prevent formation of mold, which otherwise develops on the insects in 3 or 4 days, put 15 mL (a tablespoon) of naphthalene or paradichlorobenzene or a small quantity of phenol or ethyl acetate on top of the absorbent layer. Then put a layer of cotton about 2.5 cm (1 in.) thick or a sheet of cork or cardboard on top of the absorbent layer. This separates the insects from the water, which would damage them. Drops of water may condense on the inside of the lid and fall on the specimens, especially if there is a sudden drop in temperature. To prevent this, pad the inside of the lid with cotton or put a paper towel across the mouth of the container before putting on the cover.

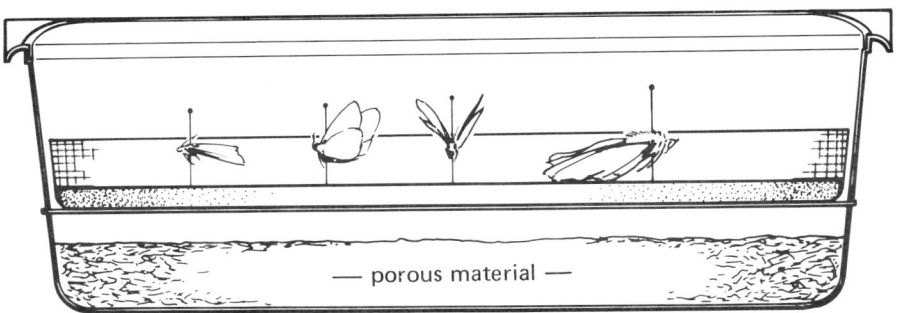

Fig. 66. A relaxing box.

Spread out the specimens that are going to be relaxed on some cork or cardboard. Do not remove papered specimens (*see* p. 82) from their envelopes, and do not pack them too tightly in the container because the moisture penetrates slowly. The time required for specimens to become relaxed depends on their size and on the temperature. At normal temperatures, small, delicate specimens relax satisfactorily overnight. Most specimens relax in about 24 hr, but very large specimens may take longer. Quick relaxing and early mounting of all insects tend to minimize discoloration and damage. Keeping the relaxing box warm helps specimens to relax quickly—even in half an hour or an hour—but this is not strongly recommended because the box needs constant attention to prevent condensation. It is best to relax specimens at a high room temperature. To keep the relative humidity in the relaxing box high, work with relaxed material at a falling room temperature. Too long in the moist atmosphere of a relaxing box discolors the specimens and eventually causes them to disintegrate. Green insects often discolor, unless you add ammonia to the relaxing box. Many of them, however, can be relaxed once satisfactorily, but become discolored if relaxed a second time. Take out one insect at a time for spreading. Finish spreading one insect before you start on another, because they dry out very quickly and in a few minutes are not safe for spreading. Do not put insects taken from a relaxing box directly on a hard surface, such as a watch glass, because dew forms at points of contact and may damage the specimens when you pick them up. Put the specimens on a piece of paper, cork, or card. The Newman relaxing tin contains a deep layer of crushed laurel leaves, which keeps fresh material relaxed and relaxes dried material.

Individual dried specimens may be relaxed with Barber's fluid, or a 30–50% solution of concentrated ammonium hydroxide. These materials are particularly useful for relaxing and rearranging the appendages. Apply the fluid to the part you want to relax with a fine brush.

Cleaning

The importance of clean specimens must be emphasized. The lepidopterous scales and other loose, extraneous debris that often adhere to insects collected in a light trap, sweep net, or dirty killing bottle may be removed by gently blowing or by careful brushing with a fine camel's-hair brush. Often, the specimens have to be washed. To clean insects such as beetles that do not have fragile wings, scales, or long hairs, rinse them briefly in a dish of ammonia, and brush them lightly if the dirt persists. Ammonia (fort) diluted with an equal quantity or more of water is recommended, but household ammonia, or even water alone, may be used. Immerse specimens with fragile wings, hairs, or scales in 30% alcohol. A good soaking may clean the specimen, but often brushing is needed. A brush suitable for use on such fragile insects may be made by cutting off all but about two dozen of the bristles of a camel's-hair brush, cutting the remaining tuft to 1 or 2 mm long, and crimping the metal to hold the tuft tightly.

To clean specimens mounted on points, dissolve the adhesive by soaking the specimen and the point in a suitable solvent. Use 30% ethyl alcohol for dissolving water-soluble adhesives, methyl alcohol for shellac, 90% ethyl alcohol for shellac gel, acetone or ethyl acetate for plastic adhesives, and xylene for Canada balsam. When the adhesive has dissolved completely, wash the specimen in 95% alcohol before transferring it to the 30%. Specimens that have become greasy because their internal organs have decomposed may be degreased by immersing them in benzene.

An ultrasonic cleaner, although expensive, is useful for cleaning dirt, grease, lepidopterous scales, and other materials from hard-bodied insects. Immerse the specimens that are to be cleaned in a solution of the cleaner, and then subject them to ultrasonic vibrations. The solution used in the cleaner can be made from a detergent, ammonia, ethyl acetate, or water, depending on whether dirt, lepidopterous scales, or grease is to be removed. Specimens stored in alcohol can usually be cleaned by placing the entire vial, including the liquid, in the cleaner. Ultrasonics are especially useful for removing fungus from specimens. Test delicate specimens, such as mites, some larvae, Hymenoptera, and others, in the cleaner before you subject them to vibrations.

After you have washed the specimens, dry them thoroughly before mounting them. If the specimen has dried in a position suitable for mounting, dehydrate it quickly by soaking it in 95% alcohol followed by absolute alcohol, and dry it on blotting paper. If it has dried in a position unsuitable for mounting, place it in water until it is sufficiently relaxed for the appendages to be rearranged. Lift it from the water on a small piece of blotting paper, dry it for a few minutes, and then place it, still on the blotting paper, in 95% and then absolute alcohol, as previously described. To pin or point specimens that have been stored in alcohol or other liquid preservatives, follow the same procedures.

It is always better for the specimens, easier for the collector, and less irritating for the taxonomists if insects are kept clean by avoiding the use of dirty killing bottles and if insects that are to be mounted dry are stored dry and not in liquids.

Temporary storage in papers

Pin medium-sized and large insects that are mounted by direct pinning through the body immediately after killing them, if possible (*see* p. 85). When immediate pinning is impractical, store such insects temporarily in paper envelopes or tubes.

Store species of such orders as Lepidoptera and Odonata, which have large wings and small bodies, in triangular paper envelopes. Each envelope is made from a rectangular piece of paper measuring from 4 × 6 to 13 × 20 cm (1-1/2 × 2-1/2 to 5 × 8 in.), folded as shown in Fig. 67. Stiff paper is best because it reduces damage from pressure when envelopes are packed together. Some collectors prefer a transparent material such as cellophane, but the insects slide around in such envelopes and they are safe to use only in a cool or dry climate. In most climates, fungal or bacterial decomposition may develop in cellophane envelopes. Put only one specimen in each envelope; fold its wings over its back. Store hard-bodied insects, such as Coleoptera and Hemiptera, between layers of Cellucotton or facial tissue in rectangular paper envelopes made as shown in Fig. 68. Several specimens may be placed in each envelope.

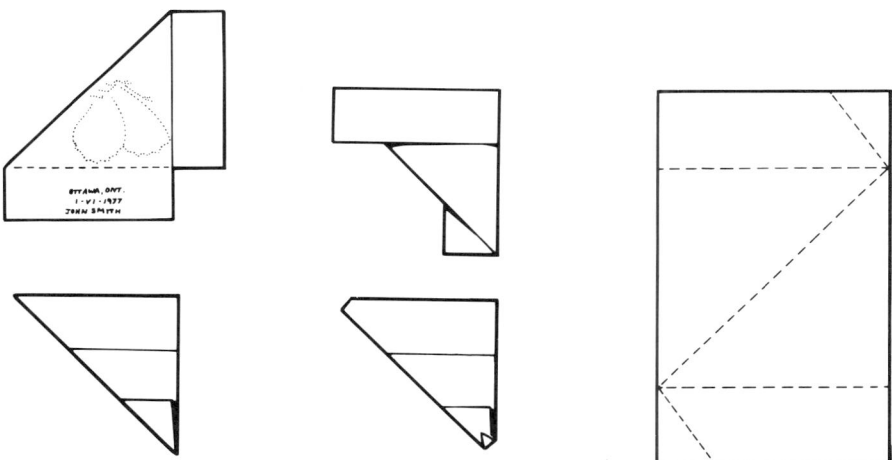

Fig. 67. Making envelopes for storing winged insects.

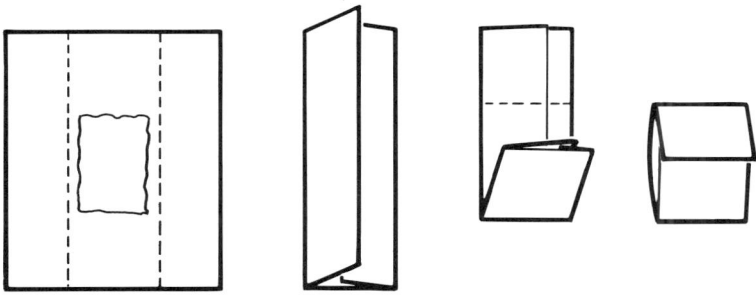

Fig. 68. Making envelopes for storing insects in layers.

Write all the pertinent data on the outside of the envelope, and for easy sorting write also the name of the species and the group to which it belongs. Write the information on the envelope before inserting the specimen, to avoid damaging the insect. A rubber stamp with interchangeable letters saves time in labeling. One of the difficulties with envelopes is that they tend to open if sorted repeatedly or stored loosely. This can be prevented by folding them in advance and storing them under pressure for a few days. Alternatively, a twist to each end of the envelope helps, but be careful not to twist the insect.

Store large hard-bodied wingless insects or insects with hard elytra in paper tubes. Make these by folding a piece of paper around the end of a pencil, as shown in Fig. 69. Again, write the data on the paper before you fold it.

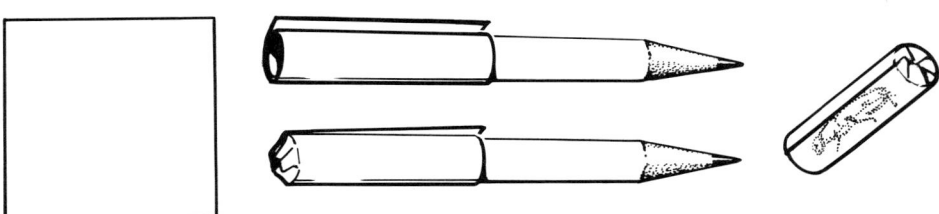

Fig. 69. Making paper cylinders for storing large-bodied insects.

Pack envelopes or cylinders containing insects with enough pressure to prevent them from sliding around but not with pressure sufficient to damage the specimens. To keep out moisture and pests be sure the box in which the envelopes are stored has a tight-fitting lid. A metal box is best. Dry the specimens thoroughly before placing them in the box, because otherwise fungal or bacterial decay may damage them. Put silica gel or a similar drying agent and naphthalene or another pesticide in the box. If the envelopes do not fill the box, fill the space with absorbent cotton. In Canada, where pests and moisture are not normally real dangers if suitable precautions are taken, perhaps the best practice is to store the envelopes in small cardboard or wooden boxes until the specimens have dried, and then these boxes may be packed in a larger pest-proof container.

Temporary storage by refrigeration

Fresh specimens may be stored for almost a week in a refrigerator provided they are kept in a tightly closed container with a leaf to prevent them from drying out. They may also be stored for longer periods in airtight containers in a deep freeze at about $-20°C$. If the container is not sealed tight, freeze-drying occurs. Line the container with soft paper to protect the specimens from contact with moisture condensing on its sides. When you remove specimens from the freezer, allow them to reach room temperature before you open the container to prevent wetting due to condensation.

Temporary storage by layering

Layering is the most widely used and satisfactory method of temporarily storing most small insects that are to be mounted dry, when it is not possible to mount them before they have dried. Pack the specimens in pill boxes or other small containers between layers of a soft material such as facial tissue, glazed cotton, or Cellucotton. Metal boxes are not recommended because of the danger of internal condensation and the development of molds. Do not use absorbent cotton, because the appendages may catch in it and break. The layers of tissue should be slightly larger than the box so that they are slightly rucked up at the edges. This holds them in place and prevents the insects from getting mixed up with the other layers. Place several layers of tissue or cotton between each layer of insects (Fig. 70). Spread out the insects on a layer, not touching one another. Pack the contents of the box with enough pressure to prevent the insects from sliding around, but not with pressure that might damage them. Suitable pressure can be achieved by packing the box with absorbent cotton over the uppermost layer of tissue, so that no movement of the insects occurs when the box is closed. Print the data about the specimens on the lid. Do not put specimens with different data in the same box, because they could get mixed up.

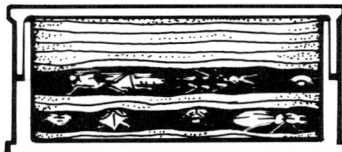

Fig. 70. A pill box containing layers of insects.

Direct pinning

Hard-bodied insects may be mounted on pins in one of three ways depending on their size, structure, and the group to which they belong. These ways are: direct pinning (Figs. 71–82), in which you insert the pin directly through the body of the insect; pointing, in which you attach the insect with adhesive to the tip of a card triangle, which is transfixed by the pin; and attaching the insect directly to the side of the pin with adhesive. Too many collectors tend to pin directly all specimens whose bodies are large enough to take pins. This is often inadvisable because direct pinning may damage or distort the insect, and, if the specimen is small in relation to the size of the pin, damage or distortion may be so extensive that identification becomes difficult or impossible. A safe rule to follow is to pin directly only those specimens (other than Lepidoptera) that obviously are too large to be mounted in other ways (*see* "Applying the Methods," p. 124, for more details).

85

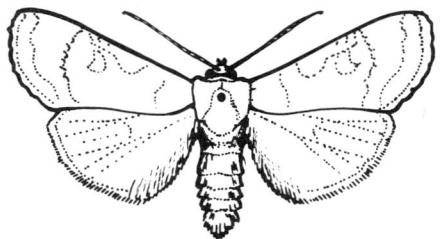

Fig. 71.   Direct pinning Lepidoptera.

Fig. 75.   Direct pinning Hymenoptera.

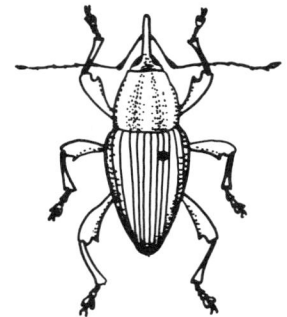

Fig. 72.   Direct pinning Coleoptera.

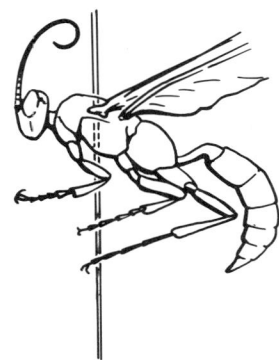

Fig. 76.   Direct pinning Hymenoptera.

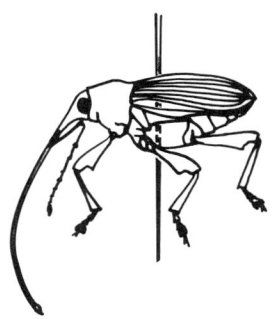

Fig. 73.   Direct pinning Coleoptera.

Fig. 77.   Direct pinning Hymenoptera.

Fig. 74.   Direct pinning Diptera.

Fig. 78.   Direct pinning Hymenoptera.

Fig. 79.   Direct pinning Homoptera.

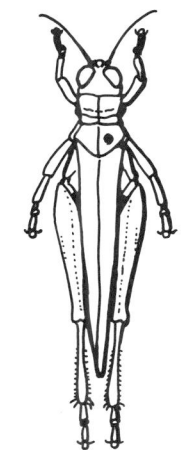

Fig. 81.   Direct pinning Orthoptera.

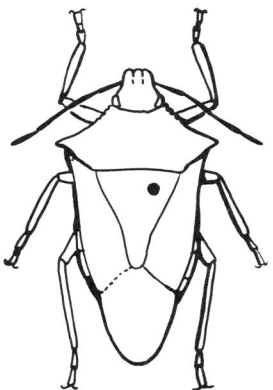

Fig. 80.   Direct pinning Hemiptera.

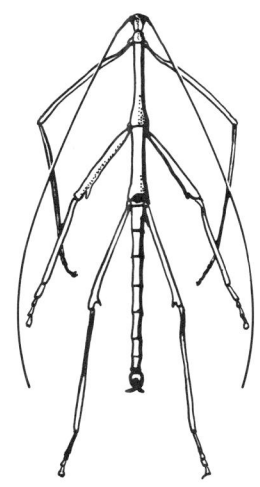

Fig. 82.   Direct pinning Phasmida.

Always use special entomological pins for mounting insects. The standard pins are made of steel, enameled or japanned black. They are of various diameters, numbered 000 (very slender), 00, 0, and 1 to 5 (very stout). The most useful sizes are 0 to 3. In theory, pins are all the same length, about 4 cm (1-1/2 in.), but in practice the lengths vary slightly from one manufacturer to another and even within one packet. Before you order pins, obtain samples, partly because pins with the same numbers from different countries sometimes differ in diameter (one kind of pins made in Czechoslovakia is a size smaller than a German brand having the same number) and partly so that you can test the samples. Watch for points that bend over easily, because of too-fine taper or too-soft metal; points that are too blunt;

heads that come off easily; lack of spring, or temper; and poor japanning, which can be tested for rust by leaving the pins in a saturated container for 24 hr. Minuten nadeln are very fine, short, headless pins of stainless steel used for very small specimens that are to be double-mounted. Elbow pins are not recommended for insects of any group; they are awkward to handle in examining specimens and the specimens work loose easily and are likely to be damaged accidentally.

To pin an insect, use the thickest size of pin that the insect can take without significant damage. It is very important to insert the pin at the correct point in its body, depending on the structure of the specimen and the different taxonomic groups (*see* "Applying the Methods," p.124, for more details). Specimens must be relaxed; if not, pinning is difficult and appendages may be broken. Moreover, the appendages must be arranged so that they and other parts of the body are displayed to the best advantage for study; for this reason, some insects must be spread (p. 90). Others must have their legs arranged so that they do not obscure the body or project in positions where they are likely to break. The antennae should be stretched out from the head, and the wings arranged so that their venation is visible and so that they do not obscure the body. A needle mounted in a handle, an insect pin, a fine forceps, or a fine camel's-hair brush may be used to arrange the appendages.

Freshly killed material is easier to pin and handle than material that has dried out and later been relaxed, because appendages break off relaxed specimens more easily. Insert the pin vertically through the body or sloping very slightly so that the front of the body is very slightly raised. Push the specimen up the pin until its back is about 1 cm (1/2 in.) from the top. This leaves enough space above the insect for the pin to be grasped with your fingers or with pinning forceps and below for data and other labels. Specimens at a uniform height are easy to examine and compare with one another, and the uniformity adds greatly to the appearance of the collection. To prevent the abdomen from drooping, pin large, heavy-bodied specimens to a vertical sheet of cork until they are dry, or store the insect box vertically.

Items of apparatus that are useful when pinning are: fine-pointed, straight, or curved forceps for handling unpinned specimens (Fig. 83); a pair of pinning forceps (Fig. 84), which is used instead of your fingers for handling pinned specimens, especially those on slender pins; a thick piece of cork; a piece of balsa wood cut across the grain, or other soft, firm material into which the insects may be pinned temporarily before labeling or transferring them to storage boxes; and a hardwood or plastic block with holes of different depths, which enable points or labels to be placed on pins at uniform heights (Fig. 85).

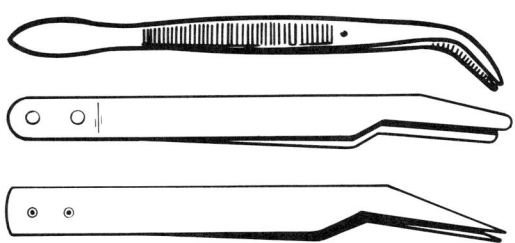

Fig. 83. Various types of forceps for handling specimens.

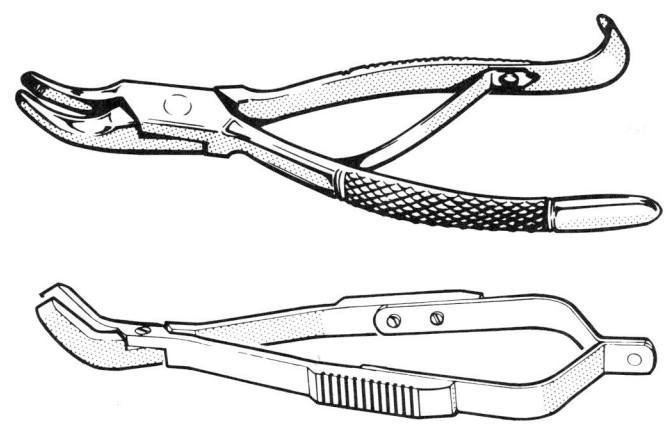

Fig. 84.
Pinning forceps for handling pinned specimens.

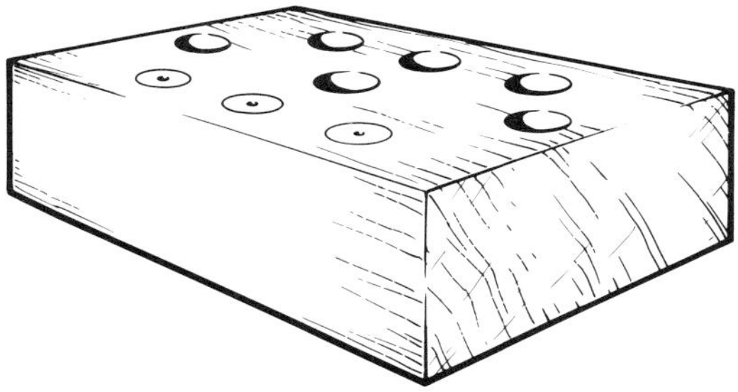

Fig. 85.
Design and construction of a pinning block.

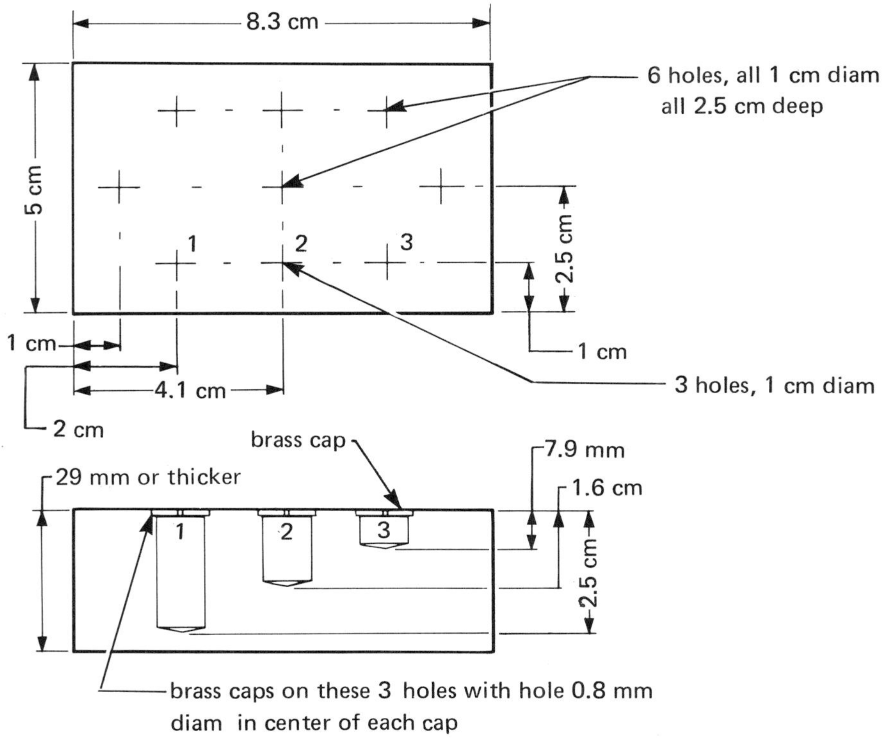

8.3 cm
6 holes, all 1 cm diam
all 2.5 cm deep
5 cm
1
2
3
2.5 cm
1 cm
1 cm
4.1 cm
2 cm
3 holes, 1 cm diam
brass cap
7.9 mm
29 mm or thicker
1.6 cm
2.5 cm
brass caps on these 3 holes with hole 0.8 mm
diam in center of each cap

Double mounting

Minuten pins are too short and weak to be handled or pinned into storage boxes without risk of damage to the specimens. To avoid these difficulties, double-mount or stage. Pin the specimen to a short strip of a soft, firm material and pass an ordinary entomological pin (No. 3) through one end of it (Fig. 86). Handle the specimen by the large pin and attach the data and other labels to it. The most satisfactory material for staging is a species of *Polyporus,* a bracket fungus. It is used extensively in Europe, and can be obtained in strips of suitable size from entomological equipment dealers in England. It is of the right consistency to hold pins well and, because it is pure white, it improves the appearance of the mounts. Alternatively, strips of cork, pith, or balsa wood may be used. Some entomologists dislike double mounts because the specimens are awkward to examine and the mounts take up more space in the collection. They are rarely used in the Canadian National Collection.

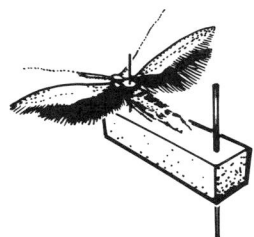

Fig. 86. A double-mount, Microlepidoptera.

Spreading

The wings of insects usually show important taxonomic characters. Insects often die with their wings closed or folded, so that these and other characters are not visible and rapid comparisons among numbers of specimens are not easily made. To overcome these difficulties, the wings should be arranged so that they are flat, horizontal, and symmetrical, and then fixed in this position. This process is known as spreading or setting.

The equipment needed consists of several spreading boards of various sizes, a number of strips of strong smooth paper, cellophane, or tracing cloth, and a quantity of fine pins. The pins should include some of the slenderest—No. 000.

A spreading board consists of two flat, parallel pieces of softwood with a cork-lined groove between their inner edges. The standard type is shown in Fig. 87. Such boards may be obtained from entomological equipment dealers, or they can be easily constructed. The flat pieces may be of cork covered with smooth paper or of a softwood that can take and hold fine pins; California redwood, pine, and basswood are excellent. If the upper surface of the board is rough, smooth it with fine sandpaper to avoid catching the wings in the irregularities and becoming damaged. The board may be flat, as shown in the figure, or its upper surface may slope very slightly up from the middle to the sides. The sloping compensates for the slight drooping of the wings that may occur when specimens are removed from the board. The cork or cloth strips in the groove should be soft enough to allow fine entomological pins to pass through them easily; if they are too hard the boards are difficult to use and the pins may bend. Spreading boards vary in the widths of the groove and the size of the upper surfaces. The height remains the same, because entomological pins are about the same length, irrespective of their thickness. An adjustable spreading board is available commercially that may be useful to a collector who spreads only a few specimens, but most collectors require several boards of various sizes.

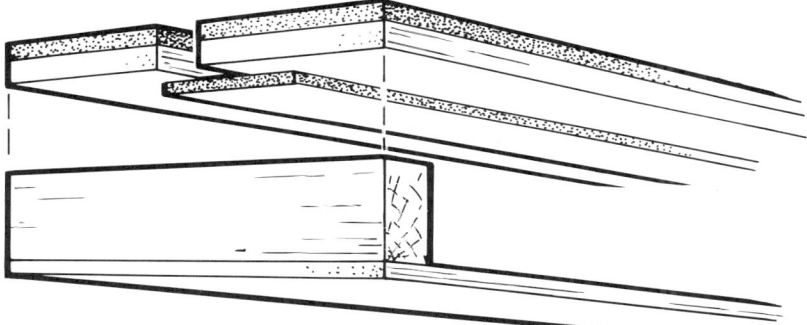

Fig. 87. Design and construction of a spreading board.

The process of spreading is easy in principle. Practice is needed, however, before it can be done easily and neatly. Specimens to be spread must be in a thoroughly relaxed condition. Select a board that is suitable for the size of the insects to be spread (Fig. 88). Make the groove just wide enough to fit the insect's body; otherwise, it is difficult to get the wings into position without damage or distortion, and the board must be wide enough that the wings do not project beyond the sides. Thrust the pin vertically into the groove so that the body of the insect rests in line with the groove and partly in it, and the wings are exactly on the top surface of the board and rest on it partly extended. Then advance the wings one at a time into the required position. This is done by inserting the point of a very fine pin (preferably a No. 000) immediately behind a main longitudinal vein of the fore wing and by pulling the wing forward over the surface of the board. If the pin is not inserted directly behind the vein, the wing may tear. In most insects, including all Lepidoptera, push the fore wings forward until their posterior margins are in line with each other and at right angles to the body. Next, bring the hind wing forward in the same manner until its anterior (costal) margin is just underneath the posterior margin of the fore wing. This may be hard to do in some insects, but the difficulty can be overcome by sticking a pin into the board at an oblique angle so as to raise the posterior part of the fore wing and to permit the hind wing to slide under. Either of two methods may be used to prevent the wings from returning to their original position. The pins holding the wings can be thrust directly into the board. This method has the advantage that it is simple and quick, but, if it is not carefully done, it may leave unsightly holes in the wings. Or, the wings can be secured by a narrow strip of waxed paper pinned to the board directly beside the groove. First, pin the paper at the front of the wings, hold the paper in your hand until the wings are in their proper position, and then pin the paper to the board behind the wings in order to press on them and hold them in place.

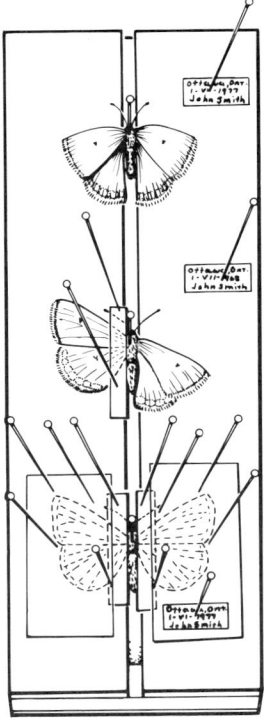

Fig. 88. Method of spreading.

A special spreading board designed to take advantage of static-electric attraction is used for small moths. The board (Fig. 89) is made of a solid piece of 2.5 cm lumber, 6.7 cm wide, and 30 cm long (1 × 2-1/4 × 12 in.). Rabbet the top along each side to a width of 13 mm (1/2 in.) and a depth of 3 mm (1/8 in.). Glue a strip of cork in the rabbet on each side and another strip of cork to the bottom, flush with all edges of the board. Glue a strip of Plexiglas (or Lucite), 2.5 cm (1 in.) wide and 1.6 mm (1/16 in.) thick, to the top of the nonrabbeted center section of the board. Before it is glued onto the board, saw the plastic strip for three-quarters of its width at 2.5 cm (1 in.). The saw kerf should be less than 1 mm (1/32 in.) wide. After the plastic has been glued to the board, drill holes through the wood in the center of each saw kerf, but not through the cork lining the bottom. These holes should be almost as wide as the saw kerf; they allow the insect pin to be inserted in the cork at the bottom of the board.

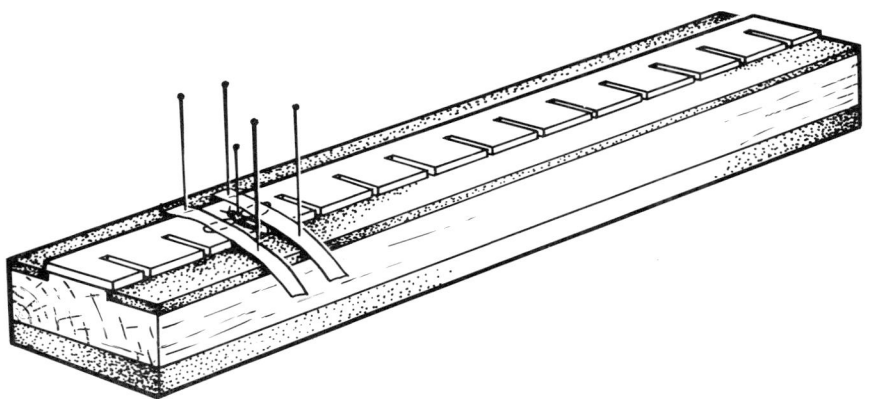

Fig. 89. A spreading board for Microlepidoptera.

To spread a small moth, slightly anesthetize it in a glass killing tube of acetic ether and then immediately pin it through the center of the thorax with a No. 000 or minuten pin. Use a microscope for greater accuracy. It is important to insert the pin perpendicular to the longitudinal axis of the moth's body. Because the moth is not dead, the muscles are relaxed, and by gently blowing behind the wings you can move them until they are semispread. Then insert the pin through the hole and into the cork at the bottom of the board. Because the moth is alive, static electricity causes its wings and antennae to stick to the plastic. Manipulate the wings and antennae into the correct position by using a No. 000 insect pin. Do not insert the point of the pin in the wings, but use it to push the wings into place by inserting it under the trailing edge of each wing near the body. When the wings are in the right position, place a strip of thin cellophane over them on each side of the body, and pin the ends of each strip to the cork at the sides of the board. Now kill the moth by placing the open end of an uncorked killing tube containing ethyl acetate over the moth for a minute or two until the moth dies. The static-electric attraction is gone by this time, but the cellophane strips hold the wings in place until the moth is dry and can be removed from the board.

This technique is particularly suitable for narrow-winged moths with long fringes, such as Gracillariidae, Lyonetiidae, Coleophoridae, Nepticulidae, and Tischeriidae. This method is safer but slower. Whichever method you use, the next step is to fix a broad strip of paper, cellophane, or tracing cloth over the outer edge of each pair of wings; otherwise, the wings may curl up and warp during the drying process. The strips are held by pins placed so that they do not pass through the wings. The antennae are arranged symmetrically, with the aid of a needle, and held in position with pins. Support the abdomen by crossing pins beneath it, if it shows a tendency to droop; or, set the board vertically on end, so that the abdomen hangs down in the correct position. Print the data belonging to the specimen on the strip or on a label, and pin it to the board beside the specimen.

Put the spreading board away in a dry, dust-free, airy place until the insects on it have dried out thoroughly. The time insects take to dry depends on their size and on the temperature and humidity. Freshly killed specimens dry in about 3 weeks, or longer if they are large-bodied; relaxed specimens normally dry in from a few days to a week or so. If the insects are taken off the boards before they have dried thoroughly, the wings may rise or droop. This may also happen if they are taken off in wet or humid weather. Drying may be accelerated by heating the specimens in a slow oven; mild heat does not harm them. It is safer to leave insects on the spreading boards too long than to risk taking them off too soon. When you remove the insects from the boards, be careful not to break their antennae or legs.

Adhesives and pointing

The choice of a suitable adhesive for attaching specimens to points or pins is important. The adhesives commonly used have various advantages and disadvantages. The adhesive must be one that binds firmly, is fairly thin so that it can be used in minute quantities, and does not pull out into threads.

Perhaps the most satisfactory adhesive widely used for small and medium-sized insects is shellac gel. It has the advantages that minute quantities bind firmly, it is thick enough to hold the specimen in place as it dries, and it stays soft for long enough that none of the common difficulties in mounting caused by rapid hardening of the adhesive occur and the specimen can be arranged in a suitable position on the point or pin at leisure while mounting is being carried out. A glass vial of the type shown in Fig. 90 is useful for mounting with shellac gel. Dip the needle affixed to the cork into the gel and the amount that sticks to it should be enough to do about 20 min of mounting. Because the gel gradually thickens as the solvent evaporates, it is advisable to mount small specimens first, and larger specimens toward the end of the 20 min. After a while the gel in the jar becomes tacky, and you can thin it by adding a few drops of polyvinyl alcohol or 95% ethyl alcohol.

Fig. 90. A screw-capped vial for shellac gel.

Pure white shellac is a good adhesive that, like shellac gel, has the advantage that only a minute quantity is needed to attach a specimen and it does not shrink as it dries. However, it tends to become runny under hot humid conditions, it cannot be used for mounting specimens that have been preserved in methyl alcohol, and beginners may dilute it too much. If it is too liquid, concentrate it before using it by pouring a small quantity into a Petri dish and leaving it until it has reached the consistency of glue, then store it in a tightly stoppered bottle or vial. A few drops of methyl alcohol added from time to time keep it from drying out.

Cellulose and other clear plastic adhesives have no advantages over other adhesives and have some serious disadvantages. The transparency is of no advantage; the solvent evaporates so fast that a skin forms as soon as the container is opened, making mounting difficult; and the adhesive shrinks quite a lot as it dries. Moreover, such adhesives do not adhere to wet or oily specimens. However, some newly developed plastic adhesives may be better than the ones we have used in the past.

Pointing probably is the most widely used method of mounting small dried insects (except Diptera). As mentioned in the description of direct pinning, insects that should be pointed are often pinned directly, to their detriment. (For information on when to point and when to pin directly, *see* "Applying the Methods," p. 124.) Pointing consists of attaching the specimen with adhesive to the tip of a small triangle of thin card that is transfixed by an entomological pin. This seems simple, but there are several errors that must be avoided if the specimens are to be mounted so that they can be examined and identified with the minimum of difficulty. Unfortunately, some collectors appear to be too careless or too thoughtless to handle their specimens carefully. As a result, they destroy specimens that are important to them in their work or that are of taxonomic value.

Points are made of heavy paper or thin card that is stiff and heavy, holds a pin well, and does not break when bent. The best material is two-ply Bristol board. Points made of celluloid and such materials have no advantages over card points and have some important disadvantages. The point used for most insects is a truncated triangle, 8 mm long, 2.5 mm wide at its base, and 0.7 mm wide at its apex (3/8 × 1/10 × 1/32 in.) (Fig. 91). Very small specimens—those 2.5 mm (1/10 in.) long or less—require points that are not truncated. Points are cut with a special punch, obtainable from entomological equipment dealers. The punch should be of good quality, and should cut the edges sharply and not raggedly. Points can be made by cutting a straight strip of card, 8 mm (3/10 in.) wide, with scissors or a razor blade as shown in Fig. 92, but this is inadvisable if it can be avoided because such points often tend to be crudely made and not to be uniform in size and shape.

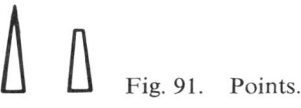
Fig. 91. Points.

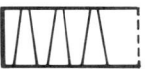
Fig. 92. A pattern for cutting points.

Insert an entomological pin (No. 3) near the broad end of the point, equidistant from the three sides, and push the point up until it is about 1 cm (1/2 in.) from the top of the pin. To keep the specimens level, use a hardwood or plastic block in which holes of the appropriate depth have been bored, or use a step block with holes bored through (Fig. 93).

Put only one point on each pin, and only one specimen on each point. The only exceptions are a pair of insects captured in copulation, in which event both may be mounted together, or polyembryonic parasites from the same host.

Because the chief purpose of mounting an insect is to prepare it for study and identification, care must be taken to ensure that no part of the specimen that may be of taxonomic significance is unnecessarily obscured by the adhesive or point. A common fault is the use of too much adhesive. In some cases, which are not uncommon, the point resembles a blob of dried adhesive from which parts of the body of the insect protrude. Use the minimum amount of adhesive that will attach the specimen firmly to the point. The amount differs with various adhesives but a little experimentation will show that, whatever the adhesive, a suitable amount is usually about one-tenth of what you think is the minimum amount. There is one part of an insect's body to which a point can be attached firmly without obscuring significant taxonomic structures. This part is the side of the thorax below the wings or margin of the tergum and above or between the bases of the legs. Although the point and adhesive will obscure one side of the thorax, the other side remains free for examination. The point or the adhesive should not extend onto the dorsal surface or onto the midventral region in most insect groups (for exceptions, *see* "Applying the Methods," p. 124), or touch the head, abdomen, or wings.

The body of the insect must be horizontal when it is mounted. Because the sides of the thorax slope inward to different degrees in several groups, the tip of the point must be bent down with forceps to a degree and length that depend on the size and shape of the specimen. For example, a specimen that has the thorax flat underneath can be mounted on a flat (unbent) point, and one that has the sides of the thorax vertical must have the tip of the point bent downward at a right angle. Most specimens require the tips of the points to be bent down to some degree less than a right angle (Fig. 94).

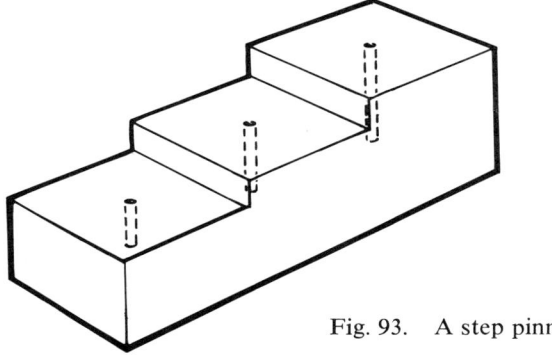

Fig. 93. A step pinning block.

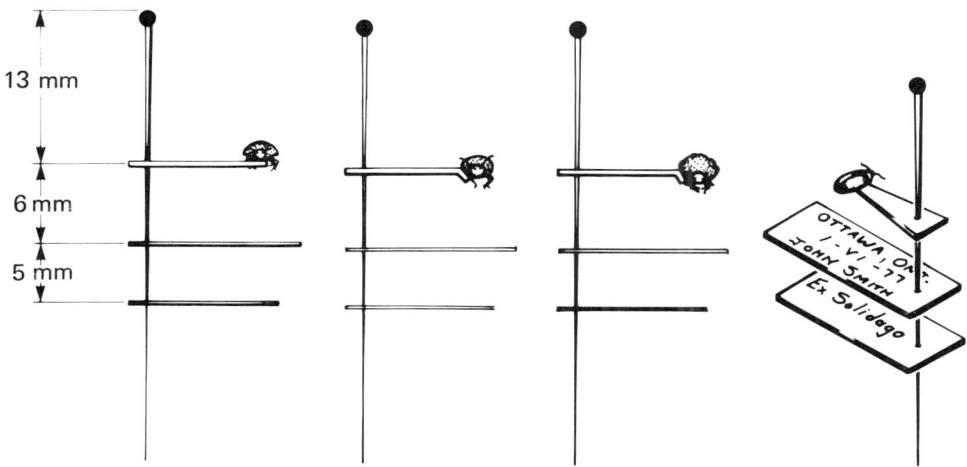

Fig. 94. Specimens on points showing positioning of points and labels and angle of specimen.

The specimens must be in a relaxed condition when they are being pointed, in order to minimize the likelihood of breaking appendages, and to enable the appendages to be arranged so that they are displayed to the best advantage for being studied or so that they are not in a position where they are liable to be broken accidentally. Specimens can be mounted more easily and are less liable to accidental damage immediately after they have been killed than if they are allowed to dry and then have to be relaxed.

A correctly pointed specimen has its body horizontal when the pin is upright, with the long axis of the body at right angles to the greatest length of the point.

To mount a specimen on a point, place the insect on its back or left side with its legs toward you. The insect may be held in this position in your fingers, or it may be placed at the edge of a small block of wood of suitable height. When pointing a series of similar specimens, some workers prefer to place a number of the insects in a longitudinal row in a shallow groove on the edge of a wooden block; the appendages of each are arranged, and then the points are applied. A piece of wood 2.5 × 2.5 × 10 cm (1 × 1 × 4 in.), with grooves of various sizes, is a useful article. Touch the tip of the point to the adhesive; press the tip against the right of the thorax; lift and invert the pin that holds the point; and adjust the specimen with your fingers or a needle. Adjusting the specimen should cause it to bind firmly to the point. A heavy specimen that rotates on the point should be straightened as the adhesive thickens but before it hardens. When you have mounted all your specimens, examine them to see that they are all attached securely to the points. Badly mounted specimens may be removed from the points by dissolving the adhesive with a suitable solvent (p. 82), followed by drying and remounting them. If you are inexperienced, practice on specimens that are of no value until you perfect the technique before you attempt to mount valuable specimens.

Preservation in liquids

Nearly all arachnids, most larval and nymphal insects, and many adult insects are soft-bodied and must be preserved in liquids that arrest decomposition, prevent shrinkage, and keep the internal organs soft enough to permit subsequent dissection. A number of preservatives have been developed for special purposes; the preservatives discussed in this publication are some of those that are most commonly used.

Arthropods may be either dead or alive when they are dropped into the preservative. The advantage of putting them in alive is that the preservative enters their bodies more easily through the spiracles and alimentary openings. The disadvantages are that the preservative often causes shrinkage and distortion of the body, and contraction of the appendages. Killing insects in hot water (p. 79) leaves them plump and limp, and also arrests the bacterial decomposition that may take place in the internal organs of large specimens faster than in preservatives. It is sometimes advisable to slit or puncture the bodies of large insects to allow the preservative to enter easier and faster.

In all cases, the preservative becomes diluted by the body fluids of the arthropods unless the volume of the preservative is 30 or more times the volume of the specimens. Therefore, drain off the preservative a day or two after the specimens have been put in, and replace it with fresh preservative. Repeat this procedure in a week or 10 days. If you do not replace the preservative, the insects will not be thoroughly preserved.

The most widely used preservative is grain (ethyl) alcohol. This is customarily used at a strength of 70–80%. Commercial ethyl alcohol contains about 95% pure ethyl alcohol; the remaining 5% is methyl alcohol and other fluids. In this publication, the alcohol referred to is 70–80% unless otherwise stated. If ethyl alcohol is not available, isopropyl alcohol may be used. The disadvantages of using alcohol are that it discolors some specimens and tends to make tissues brittle, which makes subsequent dissection difficult, especially if the alcohol used is higher than 70–80% concentration. This hardening of tissue can be minimized by adding 10% acetic acid to the alcohol. Insects in this mixture stay soft indefinitely and are easy to dissect. The mixture causes a slight swelling of soft tissue, which is usually advantageous, particularly if the genitalia are to be dissected. A 2–5% solution of glycerin added to the alcohol keeps the specimens from drying out if the alcohol evaporates. If specimens preserved in this solution are to be mounted dry, they must first be washed thoroughly in water to remove all trace of the glycerin. The shrinkage of the specimen that often occurs in alcohol can be greatly reduced by placing the specimen in 40% alcohol and slowly heating it to the boiling point, then transferring it to 70–80% alcohol.

Oudemans' fluid (*see* "Formulas," p. 169) is preferred by some workers for preserving mites, because the specimens are killed with their appendages extended, which facilitates subsequent orientation when mounting them on microslides.

AGA solution keeps insects soft and relaxed. It is useful if the external anatomy is to be studied, because it causes some distension of the specimens, so that tergal and sternal characters are shown more clearly. However, there is a tendency for the specimens to become discolored and for the internal tissues to disintegrate. Many entomologists prefer AGA to alcohol for preserving small insects.

Pempel's fluid also keeps insect specimens fairly soft and, therefore, easy to dissect. A disadvantage of Pempel's fluid, common to all solutions containing formaldehyde, is that specimens preserved in it react poorly to subsequent treatment in caustic solutions: they become brittle and the tissues do not dissolve readily. Do not use Formalin (40% formaldehyde), a common biological preservative, for insects and arachnids unless they are being prepared for histological or cytological study, or for other special purposes.

The following is a simple and good method of preserving insects, particularly large larvae, that keeps the specimens for several years in a condition suitable for dissection. Make an incision in the body of the specimens, to permit the solution to penetrate rapidly. Then place the specimen in a 5% solution of chloral hydrate in distilled water and warm it for a minute or two until it is near the boiling point. Remove the specimen and dry it on blotting paper for about 5 min, and return it to the 5% solution of chloral hydrate. After about a week, dry it again, and finally store it permanently in the solution.

If you want to preserve the specimens for histological or cytological study, or for other special purposes, see the extensive literature on biological fixatives and preservatives or consult an expert.

Keep arthropods that are in liquid preservative in glass vials, the size depending on the sizes and numbers of the specimens they contain. Each vial must contain a data label, printed or written in India ink on good-quality durable paper. If the specimens are delicate, put some absorbent cotton in with them, to prevent them from shaking about and becoming damaged. Fill the vial completely with the liquid so that no air bubbles form. Put small specimens in small vials filled with preservative, plug the vials with absorbent cotton or pith, and store them upside down in a screw-capped jar that is partly filled with preservative (Fig. 95). It is most convenient to keep only a single series of vials in a circle inside the edge of the jar, held in place by a large central plug of cotton, because if the jar is filled with vials it takes time to find the one you want. This method is recommended for specimens that are not handled often, because evaporation of the preservative is greatly reduced. Larger, neoprene-stoppered vials are convenient for specimens that are handled more often. Homeopathic vials (Fig. 96), of 10.6 mL (3-dram) capacity, 65 × 19 mm, accommodate most specimens, and their stoppers fit tightly enough to prevent evaporation. One of their disadvantages is that it is sometimes difficult to pick out the specimens. Specimens are easily picked out of shell vials, but evaporation is a serious problem. The stoppers should be of the best quality of neoprene, and air has to be bled from the vial before it can be stored. To stopper a vial that is well filled with liquid, hold a pin or fine wire beside the stopper to let the air escape. Remove the pin or wire when the stopper has been inserted to the desired depth. The system of storage using 3-dram homeopathic vials with neoprene stoppers is the most satisfactory, and collections stored this way last for many years without evaporation.

Mites and other small arthropod specimens that have been accidentally allowed to dry out can be restored by placing them in warm 50–80% lactic acid for 2 days, and then replacing them in alcohol.

Fig. 95. A screw-capped jar containing vials of specimens.

Fig. 96. A neoprene-stoppered vial containing specimens.

Dry preservation of soft-bodied insects

There are two common methods of dry-mounting soft-bodied insects and related animals that usually are preserved in liquid preservatives. Specimens preserved in this way are useful for displaying, but they are inferior in most respects to specimens in liquids for studying. The most widely used method is to inflate the empty skins with air or wax and to dry them; other methods involve replacing the natural moisture with a liquid that hardens the tissues so that little shrinkage occurs and decomposition is arrested, and embedding in transparent plastic. Only the first method is described here.

Caterpillars and other larvae of similar form and structure can be inflated. Inflated larvae hold their colors (except greens) better than specimens preserved in fluids, but often the setae break off and, of course, the internal organs are missing. The principle is to blow up the empty skin like a balloon and to keep it in this condition in an oven until it has dried hard. The following equipment (Fig. 97) is needed: a piece of glass tubing drawn out at one end to a narrow aperture (or a natural straw); a pair of clips made from lengths of watch spring or hair clips (bobby pins) attached to the narrow end of the glass tube and bound to the tube with wire; a small oven, which may be made from an empty can, that has one end removed and an opening near the top of the other end, placed on its side over an alcohol lamp; a round pencil; pins; pieces of straw or soft wire; and adhesive.

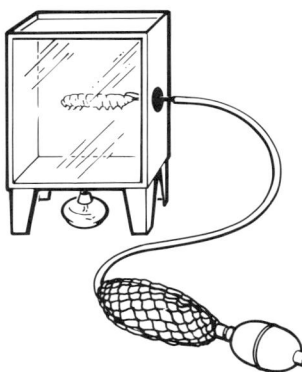

Fig. 97. An apparatus for inflating larval skins.

The method of inflation follows. Kill the larva in hot water and stretch it on a sheet of blotting paper. Make an incision around the anus with scissors or a sharp razor, freeing the intestine. Then force the contents of the body out through the anal opening by rolling the pencil the length of the body, starting immediately behind the head capsule. Next, place the tapered end of the glass tube in the aperture through which the viscera are removed and hold the empty skin to the tube by the clips. Now inflate the skin by blowing steadily through the tube, either from your mouth or by a double-action constant-pressure rubber bulb, and simultaneously hold the skin in the oven. In the oven, it dries quickly in the inflated position, which, if the air pressure is not too great, is the natural position of the larva. The oven should not be too hot; otherwise the skin may scorch and colors may be lost. When the skin has dried, remove it from the oven, detach it from the tube, and mount it. Alternatively, the skin may be inflated with melted beeswax, or equal parts of beeswax and paraffin wax, by means of a hypodermic syringe. Specimens treated in this way are less fragile, and the method has the additional advantage that the wax may be tinted with dye of the same color as that of the living larva. A skin may be mounted in either of two ways: by attaching its feet with adhesive to a length of straw held at one end by an entomological pin; or by inserting into the anal opening a tiny sharpened cork coated with adhesive and attached to a piece of soft wire, the other end of the wire being wound around the pin. Alternatively, the cork is pierced by the pin. Because the incision at the anal region of the larva may destroy some characters of that region, it is advisable to inflate a few larvae of each species from the head end as well.

The lost colors of inflated larval skins or of chemically dehydrated specimens may be reproduced with paints, provided you know what and where these colors were. Color photographs of the living insects are a great help in this. The best material for study, however, is specimens preserved in fluids together with color photographs of them alive.

Freeze-drying of soft-bodied arthropods and immature stages is a method that has been developed and is extremely useful for display purposes. Specimens retain their color and are natural and lifelike. The method involves the vacuum dehydration of specimens at below freezing temperatures. For further information, see the reference list for methods of freeze-drying. Similar results may be obtained by placing fresh pinned specimens in a freezer for long periods to allow them to dry.

Microscopical preparations

Insects Many special techniques for preparing, staining, and mounting insects or parts of insects for microscopical study have been developed. This publication describes only a few methods: some that are in common usage for general collecting, whose descriptions follow; and some that are used to prepare specimens of particular taxonomic groups for identification, which are described in a later section, "Applying the Methods," p. 124. Modified or alternative methods for special purposes are given here, and additional information may be obtained from the references listed at the end of this publication.

Preliminary preparation of a specimen or part of a specimen, such as the genitalia, usually involves removal of the soft parts so that the sclerotized structures and membranes are unobscured. The soft parts can be dissolved in a solution of potassium hydroxide (KOH). The usual solution is 10% in water, but a weaker solution may be better for delicate objects. Leave the object in the solution until the soft parts have dissolved; this may take from a few minutes to a day or more, depending on the size, nature, and condition of the object. A convenient method is to place the object in the solution in a flat-bottomed glass cup, 14 mm wide by 7 mm deep, or some other convenient size. Small objects may be cleared in the cavity of a hanging drop microscope slide. The process may be accelerated by heating the solution. This reduces the time to between 1 and 20 min. The solution may be heated in a glass Stender dish or in a porcelain dish on a hot plate or water bath. Do not allow the solution to boil, because boiling often damages specimens and destroys delicate ones; supply only enough heat to set up convection currents in the solution. Add water from time to time as the solution evaporates because evaporation increases the strength of the caustic, which can easily damage the specimens. The length of time needed for best results with a particular specimen can only be learned by experience; it is better to give a specimen too little treatment than too much, because the latter can cause irreparable damage.

When the soft parts have dissolved, transfer the object to water and rinse off all traces of the caustic. A piece of soft wire with a small loop at the end is useful for picking the object out of the solution. Examine the object with the aid of a dissecting microscope, and expel any liquefied soft parts that remain within it by gently pressing them with a needle. Finally, if staining is unnecessary, mount the object by using one of the methods described in the following paragraphs.

Some structures are too transparent to be readily studied, either because they are unpigmented and only weakly sclerotized in their natural condition or because they have been softened or depigmented by overtreatment in caustic. These structures should be stained. A simple stain for chitin is carbol-fuchsin. Place the object in this stain in a small dish for as long as necessary—up to 12 hr or more. Little harm is done if it is left in too long, because excess stain can be removed by transferring it to a solution of 95% alcohol containing 3% by volume of concentrated hydrochloric acid until the object reaches the desired color.

A specimen or part of a specimen may be mounted, with or without the preliminary treatments described above, in either of two ways: in a microvial or under a cover glass on a microscope slide. The advantages of a preparation in a microvial are that it can be examined from any angle, it can be dissected at any time, and it is always attached to the specimen from which it came. The advantages of a slide preparation are that it is more nearly permanent, it is less liable to accidental damage, it can be examined more rapidly, and quick comparisons among a number of slides can be made readily.

Microvials are small vials about 10–15 mm long and 4–5 mm in diam. The corks are of the best quality and cylindrical in shape. Microvials may also be made of semitransparent polyethylene and have silicone rubber stoppers, which are impervious and nonreactive to glycerin. The greater transparency of glass vials allows for better examination in situ. Put a very small drop of glycerin in the bottom of the vial with a fine hypodermic syringe. A common fault is the use of too much glycerin, which may soak into the cork and dissolve the pith, turning the glycerin brown, causing the object to become brittle, corroding the pin, and eventually destroying the data labels and even the specimen. This does not happen with polyethylene vials, because the glycerin does not creep up the sides of the vial. Use just enough glycerin to cover the object. The walls of the vial should be free from glycerin; otherwise, it may soak into the cork, leaving the preparation dry. Put the specimen into the glycerin in the vial, cork the vial, and transfix the cork with the pin of the specimen from which the object was taken (Fig. 98). When the pin is in position, the vial should tilt down slightly, to keep the glycerin at the bottom, though this is not necessary if the walls of the vial are dry. For study, uncork the vial, pick out the object with a looped wire having a pin hooked at its tip or with fine forceps, and examine it in glycerin in a watch glass or cavity slide under a microscope. Be careful when you are removing or replacing the object in the microvial not to smear the sides of the vial with glycerin, because the glycerin corrodes the cork if exposed to it. Microvials are particularly useful for preserving genitalia along with the specimen from which they were removed and for preserving larval and pupal skins along with the adults that emerged from them.

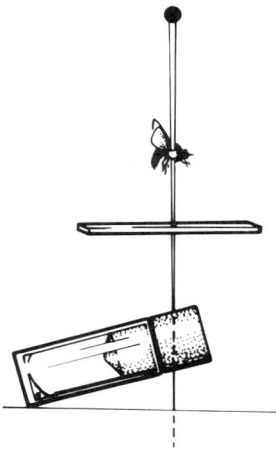

Fig. 98. A microvial containing associated parts of a pinned specimen.

A simple method of mounting genitalia (especially of beetles) for most taxonomic needs, and of keeping these organs associated with the insect, is as follows. Dissect the genitalia and immediately transfer them into distilled water and boil them for a short time to eliminate all the air bubbles. In very large and strongly sclerotized structures, such as the aedeagi of larger beetles, use a weak, about 5–10%, solution of KOH for clearing. For most insects, transfer the genitalia directly from the distilled water into dioxane for a few minutes (2–20 min), depending on the size and sclerotization of the structure, to clear it. If a solution of KOH is used for clearing, transfer the aedeagus into distilled water for about 4 or 5 min before putting it into dioxane. Place a drop of Canada balsam (the amount depending on the size of the genitalia) on the front part of a celluloid plate. Then quickly transfer the genitalia from the dioxane into this drop of Canada balsam, and with the aid of a needle manipulate it into the correct position. For this method, use thicker than usual Canada balsam so that the genitalia are kept in position until dry. Pin the celluloid plate about 6 mm (1/4 in.) under the insect so that the genitalia in the drop of Canada balsam is situated directly in front of the beetle (Fig. 99). This method is especially suitable for studying the structures of the aedeagal internal sac in Coleoptera, but it can also be used for mounting other parts (mouthparts, modified abdominal sternites, tergites, and so on).

Fig. 99. A celluloid slide mount affixed beneath the mounted specimen.

Specimens mounted on slides may not need to be dehydrated and cleared, depending on the kind of mounting medium used. Media that do not need to have the specimens dehydrated and cleared first are polyvinyl lactophenol, Hoyer's, de Faure's, Berlese's, and glycerin jelly. However, dehydrating and clearing must be done before mounting in Canada balsam or in any of the other widely used synthetic substitutes.

Polyvinyl lactophenol, Hoyer's, de Faure's, and Berlese's media are very similar; the first two are preferred by many entomologists. They are simple to use and best for specimens that are apt to distort or shrink if mounted in other media. Put a drop of the medium on a slide, place the specimen in it alive, and the insect dies with its body and appendages extended; or you can put it in after it has been killed, treated with caustic, and washed in water. Apply a glass cover. If you are going to use Berlese's medium, warm the slide gently for a few minutes and then leave it a week or so to dry. Specimens may be mounted in Hoyer's medium direct from alcohol. To mount in glycerin jelly, put a small piece of the jelly on the slide, and heat it until the jelly melts. Place the specimen in the liquefied jelly and apply the cover glass before the jelly solidifies. A disadvantage of glycerin jelly is that air bubbles get trapped under the cover glass. Preparations in these media or in glycerin jelly remain in good condition for months or years, but for permanency or in arid climates it is advisable to ring the cover glass to prevent evaporation of the medium. Duco enamel, Murrayite, Glyptal electrical finish, and gold size are a few of the most suitable substances for ringing.

The process of dehydration involves putting the object, after suitable preliminary treatment, in graded strengths of ethyl alcohol: 50, 70, and 90 or 95%, and absolute alcohol. The length of time in each depends on the size and nature of the object; the time varies from a minute or so to several hours and can be learned by experience. If the object has been treated with caustic, so that only the chitinous structures and membranes remain, it may be put through the alcohols fairly quickly, probably in a few minutes. For mounting in Canada balsam or any of its synthetic resinous substitutes, place the object in a clearing agent after it has been removed from the absolute alcohol. However, if it has not been fully dehydrated, the clearing agent becomes milky. If this occurs, return it to the absolute alcohol for a short time. Clearing agents commonly used are xylol, clove oil, oil of wintergreen, and cedarwood oil. Leave the object in the clearing agent until it clears (becomes transparent or translucent), then mount it on a slide (Fig. 100) in a drop of the mounting medium. Do not leave the specimen in the clearing agent any longer than is necessary because the specimen may shrink. For example, fragile Microhymenoptera require only from 30 sec to 1 min in the clearing agent. Tough chitinous objects that have been treated in caustic may be put directly into 95% alcohol without going through the weaker solutions, but the direct alcohol treatment often shrinks soft, delicate objects.

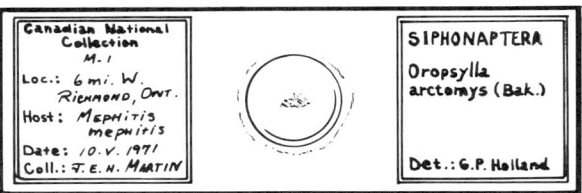

Fig. 100. A slide mount showing position of specimen and labels.

Large or rounded objects may become flattened and distorted by the cover glass as the specimen dries and shrinks. To prevent this, put a hard object under the cover glass or use slides with hollow-ground cavities. However, the latter are expensive. A simple and satisfactory way is to put a ring of a card of suitable thickness under the cover glass (Fig. 101). Cut out the rings with the use of two cork borers, one the size of the cover glass, the other smaller. To make the rings, first cut out the central disk with the smaller borer; then cut the ring from the card with the larger borer. Attach the ring firmly to a slide with a mounting medium, and allow it to dry before you use the slide. To mount, fill the ring with the mounting medium. Use more mounting medium than fills the ring, to allow for subsequent shrinking when it dries. Put the object in the medium within the ring, and apply the cover glass.

Be sure the mounting medium is dry before you handle the slides, otherwise the medium, the object, and the cover glass may flow off and be damaged or lost if you tilt the slide. Drying is particularly important with

Canada balsam and other nonwater-soluble media; use one of the ovens designed for this purpose. You can make an adequate hot plate for drying slides by putting a sheet of glass over the open top of a box that contains two electric light bulbs. The heat given off by the bulbs is sufficient to dry the slides fast enough to prevent the formation of bubbles in the mounting medium. Always handle slides by their edges, not by their broad surfaces.

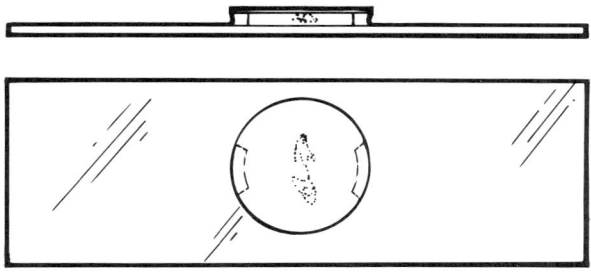

Fig. 101.
A supported cover slip for mounting large specimens or appendages.

Mites For mites, the preparation of permanent resinous mounts involving treatment with caustics, dehydration, clearing, and sometimes staining is not recommended. The procedures are time-consuming, and each step requires some manipulation of the minute and usually fragile specimens. Distortion of delicate parts sometimes occurs during drying, and remounting is a hazardous operation.

Temporary preparations and semipermanent water-soluble mounting media are the two preferred methods for studying mites on microscope slides. Temporary mounts are particularly useful for studying heavily sclerotized, bulky-bodied, or large mites. Lactic acid is used as a clearing fluid and as a temporary mounting medium. Living or preserved mites can be placed in cold or warm lactic acid, varying in concentration from 50–70% for softer-bodied Prostigmata and Astigmata to 80–100% for heavily plated Mesostigmata and Cryptostigmata (Oribatei). Good clearing results are also obtained by using lactophenol or Nesbitt's fluid (see "Formulas," p. 169), which clear more heavily sclerotized specimens faster and more effectively. Nesbitt's fluid is riskier to use than lactic acid because specimens can disintegrate if left in it too long.

To study the cleared specimen, place it in a drop of lactic acid on a slide, and cover it with a cover slip. For larger mites use a cavity slide to avoid distorting the specimen by the pressure of the cover slip. Gently heat the preparation on a warming plate until the desired degree of clearing is achieved. Too rapid heating may rupture specimens. However, for engorged blood-sucking mites you may need to pierce the body cuticle with a fine needle to hasten clearing. Open-cavity slide preparations, in which a square cover slip only partly covers the cavity that is filled with lactic acid, are satisfactory for studying larger mites with moderate, "dry" objectives. The specimen can be oriented with the use of a fine needle or by slightly moving the cover slip (Fig. 102). Closed-cavity slide preparations are often more convenient for studying smaller mites with oil immersion objectives, because small changes in position of the specimen can be effected by slightly moving the cover slip.

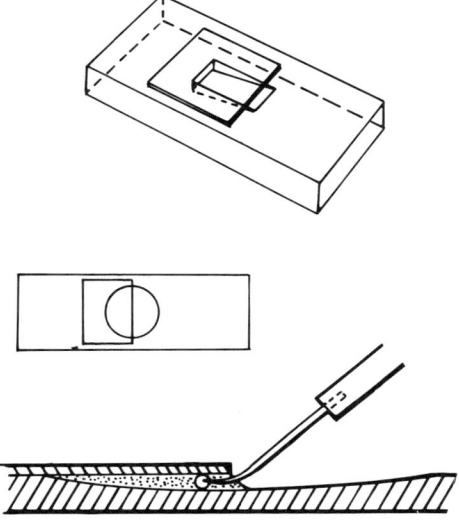

Fig. 102.
Cavity slides for large specimens and showing the method of orienting the specimen.

A wedge-shaped cavity in a slide is the most satisfactory for general use, because it accommodates large or small specimens in the deeper or shallower parts of the depression. Preparations in lactic acid on cavity slides may be sealed with Glyceel, Gyptal, or another suitable substance, and maintained for months or even a few years in this condition. However, the specimens are usually returned to vials of alcohol for storage.

The advantages of temporary preparations are that the specimens can be turned in any position for studying all the structures, and that they are not subject to injury from the deterioration of a mounting medium. The disadvantages of this method are that it is tedious, and it requires excessive direct handling and manipulating of the specimen and much more storage space.

A variety of semipermanent, water-soluble mounting media are used for mounting mites, particularly the gum-chloral or so-called "modified Berlese media." Of these media, Hoyer's medium (*see* "Formulas," p. 169) is the most widely used and satisfactory for clearing specimens; it has a suitable index of refraction, and is convenient, practical, and permanent. Living mites from land or from water, and specimens preserved dry or in alcohol can be mounted directly into Hoyer's medium. Specimens that have been cleared in lactic acid or lactophenol should be washed before you mount them, to avoid causing crystallization in Hoyer's medium. If you use Nesbitt's fluid as the clearing agent, the washing step is unnecessary. If you do your mounting directly in Hoyer's medium, apply a cover slip, and heat the slide preparation in an oven or on a slide warmer at 40–50°C for 2–4 days, depending on the degree of clearing action needed from the chloral hydrate ingredient in the medium. This heating also hastens drying and setting of the preparation. However, heat-treated slides must be left at room temperature for another week to allow further air-drying and for the cover slip to flatten again after buckling and becoming slightly convex during oven-drying.

A disadvantage of gum-chloral media is their impermanence as affected by atmospheric moisture. They are hygroscopic and may take up water from very humid atmospheres, but more seriously, under the very dry conditions that occur in north temperate areas, they lose water, and air is drawn under the cover slip in a fine pattern resembling crystallization. To prevent or greatly reduce this impermanence, thoroughly dried preparations should be ringed with a water-insoluble compound such as Glyceel or Zut paint derivative, Glyptal paint, Canada balsam, or euparal. If slides are ringed before the medium is dry and well set, stresses still acting on the cover slip may crack the ringing seal and allow air incursion again. If a Hoyer's preparation deteriorates as a result of inadequate washing or poor ringing, retrieve the specimen by scratching or dissolving away the ringing compound and soaking the slide in warm water. The cover slip can then be gently removed and the freed specimen remounted directly in Hoyer's medium on a new slide.

A number of other semipermanent and permanent mounting techniques are used, particularly with the use of glycerin jelly, C-M medium (methylcellulose fluid), and glycerol.

In making a slide mount by any method it is important to arrange the object on the slide so that it is in a position suitable for study and customary for objects of its nature, that appendages or other structures are spread out and displayed to the best advantage, and that similar objects are arranged in the same way, to facilitate comparisons. The arranging may be done with needles when the object is in the mounting medium on the slide but before the cover glass is applied. Each slide should bear labels giving the necessary data (*see* "Data Labels," which follows) and in addition the identity of the medium in which the object is mounted. The latter is important because it indicates the solvent to be used—water for polyvinyl lactophenol, de Faure's, Hoyer's, or Berlese's; alcohol for euparal; xylol for Canada balsam—if the object has to be removed from the slide for remounting or further dissection.

In any work that involves removing parts of specimens, take care not to mix up the parts or associate any part incorrectly, either directly or by cross-references on the labels, with the specimen from which it was taken.

Data labels

Adequate labeling is one of the most important aspects of collecting and preserving insects and arachnids. A mounted or preserved specimen is a source of information in two ways: some of the information can be obtained by examining the specimen itself; the remainder is given on the label or labels attached to or associated with the specimen. Both kinds of information must be available. The information provided by the specimen should be as complete as possible if the arthropod is undamaged and is mounted or preserved carefully in the approved fashion. The label or labels (Fig. 103) should provide the history of the specimen.

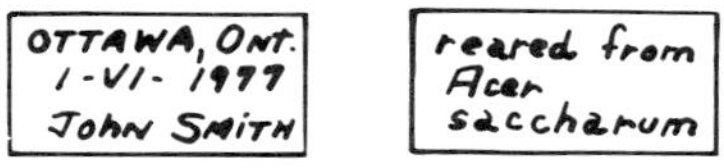

```
┌─────────────┐  ┌─────────────┐
│ OTTAWA, Ont.│  │ reared from │
│ 1-VI- 1977  │  │ Acer        │
│ John Smith  │  │ saccharum   │
└─────────────┘  └─────────────┘
```
Fig. 103. Data labels.

All information on labels must be machine-printed or hand-printed; information other than dates should not be indicated by letters, numbers, cryptic abbreviations, or other esoteric symbols when it can be avoided, because such unintelligible data cannot be interpreted without the key, which may be unavailable to others, or lost irretrievably if the key is lost.

Labels for specimens should be of good-quality stiff paper of 100% rag content. Machine-printed labels are the most satisfactory; 3-1/2 or 4 points are large enough sizes. Labels printed by the offset process can be reduced to about these type sizes. Collectors who regularly work in the same locality find it convenient to have labels printed ahead that give the name of the locality and the collector, and leave space for appropriate dates to be completed by hand. If not machine-printed, the labels should be hand-printed clearly and legibly in India ink with a fine pen.

Labels for pinned specimens should be as small as possible and still remain legible. They should not be larger than 6 × 16 mm (1/4 × 5/8 in.) with three or four lines of writing or type. Print the locality and date of capture and the collector's name on one label, and additional information, when available, on another. Every specimen must have its own label or labels; it is not sufficient merely to affix labels to one or a few of a number of specimens that have the same data. Pin the label with the specimen or its mount. The height of the label on the pin varies with the group. For insects having short legs that are liable to break, put the label about 16 mm (5/8 in.) above the point of the pin. If there is a second label, place it 5 mm (3/16 in.) below the first one. Place identification details and other labels below this one. To keep the heights of labels uniform, use a hardwood or plastic block with holes of the appropriate depths (p. 88). With a spread specimen, insert the pin through the center of the label and turn the label so that it is at right angles to the long axis of the body, that is, the top of the label is toward the insect's head. For a pinned specimen that is not spread, or a dipteron attached directly to a pin with adhesive, place the label in line with the long axis of the body and turn the top of it to the right, and insert the pin at a point in the label so that the lengths of label before and behind the pin are the same. With a pointed specimen, put the label at right angles to the long axis of the insect with the top toward the head and insert the pin in the label near the end farthest from the tip of the point. When pinning a label, be careful not to insert the pin where it would mutilate or obscure any of the information thereon; insert it between, rather than in the middle of, words or figures.

Use the same quality of labels for specimens in liquid preservative as that used for pinned specimens and put them into the container with the arthropod. Use the best quality of India ink for printing labels to prevent the printing from running or dissolving when the label is placed in the preservative. The labels and printing should be large and clear enough to be legible through the vial so that it is not necessary to remove the label to read the data or locate a species. Do not use paper or labels of other material attached to the outside of containers, because they come off too easily.

Securely attach adhesive labels to microscope slides with their data machine-printed or clearly hand-printed in India ink. Usually, all the data associated with collecting the specimen are placed together on one label affixed to the left side of the preparation. A blank label for determination is affixed to the right side; additional information, such as the mounting medium used and identification or research code numbers, can be noted on this label also.

Label every specimen with the following information:

(*a*) The exact locality where the specimens were captured by name, if it has a name, and the geographic location so that it can be found on a small-scale map. Give the name of the province or state in which the locality is situated, and the name of the county, district, or township, or the nearest place that has a post office, or some other nearby geographical feature readily found on a map and not duplicated by this name in the same province or state. If the locality or district is unnamed, give the latitude and longitude of the place of capture, the township and range and sometimes the section, or the distance and direction from some easily located geographical feature. In hill or mountain districts, always give the altitude, which is important—often more important than the exact locality. Remember that the information on the place of capture should be clear to someone else with no personal knowledge of the region who may attempt to find the locality on a standard small-scale map, or who may attempt to collect a similar specimen at that locality.

(*b*) The date of capture or, for reared specimens, the date of emergence. Include the day of the month, the month, and the year. The day and the year are indicated by arabic numerals and the month either by an abbreviation of its name or by roman numerals. Do not use arabic numerals for the month because they cause confusion between the Canadian and American systems of writing dates. For example, the 4 August 1973 should be given as 4.Aug.73 or 4.VIII.1973. It should not be given as 4.8.73 because this could be misunderstood to mean April 8, 1973, in American terminology.

(*c*) The name of the collector. The collector's name may provide a source of additional information on specimens; it also gives credit to the collector when new or rare species are recorded; and occasionally, it provides an indication of the reliability of the data when the collector's collecting, preserving, and labeling methods are known.

The following additional information, where applicable, and whenever it is available, should be given on supplementary labels:

(*d*) The name of the host plant or animal. It is important to include how the insect or arachnid was associated with the host, as an indication of the relative value of the information as life history or ecological data. The fact that the specimen was bred from, feeding on, ovipositing on, or even resting on the host is obviously of more value than that it was beaten from or swept from the host. However, the latter are of far more value than indefinite terms, such as from, on, ex, or host, that, except for parasites, can cover a variety of associations. The full, unabbreviated name of the host should be given, when it is known. If the host has not been identified to species or genus, a general description is better than no information. Reared material always must be labeled as such, because it may differ consistently from material collected in the field.

(*e*) The type of habitat in which the specimen was collected, if host information is not available. If possible the habitat should be indicated by reference to the plant association or to the dominant plant of the situation in which the specimen was collected, such as on short-grass prairie or among spruce. If this information is unavailable, a general description of the habitat may be useful, such as in marsh or in flood debris.

(*f*) Any information noted on the habits of the insects, such as flying in sunshine, at certain flowers, or in a light trap. Be sure to note on the label if specimens were captured in copulation, if they are mounted separately.

(*g*) The mounting medium used (for specimens mounted on slides) and the preservative, if other than alcohol (for specimens preserved in liquids).

(*h*) Collection accession number, or references to additional information that is too extensive to be given on labels, or numbers associating the specimens with the preserved remains of their early stages or with microscope slides of their genitalia or other structures.

Storage and care of collections

Protect your insect collection from jarring or vibration, which may damage the specimens; from dust, which accumulates on the specimens; from light, which fades their colors rapidly; from pests, which destroy or damage the insects; and from dampness, which encourages fungal damage or bacterial decomposition and which relaxes spread specimens.

Storage and care of pinned insects

It is important when you are storing dried pinned insects to be sure that the pins are held firmly, so that they do not work loose thereby allowing the specimens to damage themselves and one another. Line the container with a substance that is resilient enough to hold them firmly. The best substance for a pinning layer is compressed cork or polyethylene foam, which is available from a dealer in entomological supplies. Some kinds of compressed cork that are not manufactured specifically for entomological purposes contain softening or binding agents that corrode pins. Good, soft balsa wood holds fine pins well, but it is less resilient than cork, and must be thicker and the pins must be pushed in deeper. In hot, humid weather, balsa wood corrodes the points of the pins. If it is used in cabinet drawers, where the jarring is largely unidirectional, with the grain parallel to the sides of the drawers, pins eventually work loose parallel to the grain. Some of the wood pulp composition materials are satisfactory, but others are not; therefore, it is wise to test samples before using them. If more suitable materials are not available, you can use corrugated cardboard. Place two sheets with their flat surfaces up and the corrugations at right angles to one another. The substance that you select for a pinning surface must be fixed firmly in the container with strong adhesive so that there is no possibility of it working loose.

Push the pins well into the pinning layer. This is particularly important if the containers are moved or handled frequently. Pinning forceps (p. 89) makes it easier to handle the pins and reduces the possibility of accidentally damaging the specimens; damage resulting from pins bending is particularly liable to occur if you handle small-gauge pins with your fingers.

Any wooden or cardboard box deep enough to take pins and having a tight-fitting lid is satisfactory for storing pinned specimens. For example, cigar boxes are satisfactory provided they are deep enough. Glass-topped cases may be used, provided they are stored in the dark when their contents are not under examination. The most suitable containers are, of course, the boxes and cabinets specially designed for the purpose. In North America, the most popular and widely used standard container is the Schmitt box (Fig. 104). It is a wooden box that has a tight-fitting lid and its floor is lined with compressed cork covered with white glazed paper. The outside measurements are 23 × 33 × 5.5 or 6 cm (9 × 13 × 2-1/4 or 2-1/2 in.). In Europe, storage boxes of various sizes that have pinning surfaces on their floors and lids are preferred. Large collections are usually stored in insect cabinets, each of which usually has 10–50 drawers. The drawers have tight-fitting, removable glass tops and are available in two standard sizes. The type used for the Canadian National Collection has drawers with outside measurements of 46 × 46 × 6 cm (18 × 18 × 2-1/2 in.); the type of drawer used at Cornell University has outside measurements of 42 × 48 × 8 cm (16-1/2 × 19 × 3 in.).

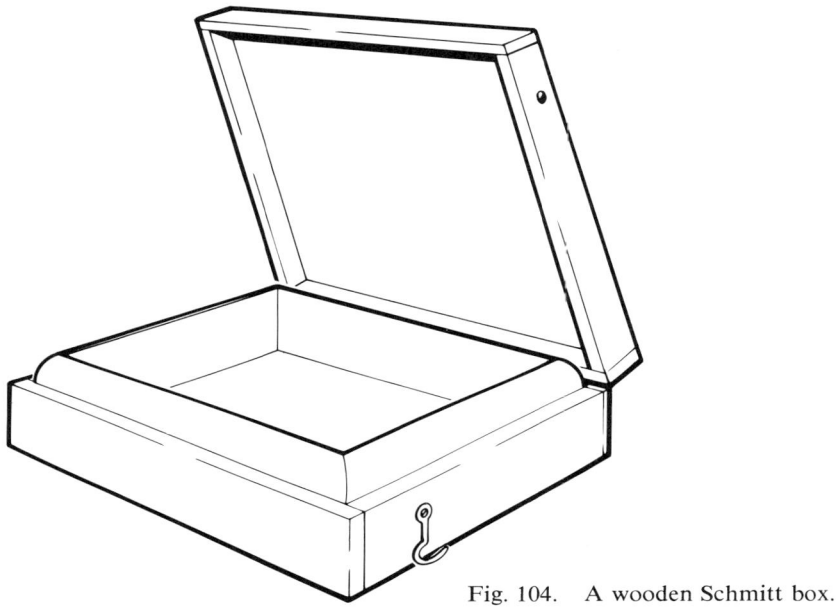

Fig. 104. A wooden Schmitt box.

For large insects line the cabinet drawers with compressed cork and cover it with white glazed paper, to which the insects may be pinned. The unit tray system is preferable for medium-sized or small insects. Specimens of each species are pinned into a cardboard tray having the bottom lined with compressed cork covered with white glazed paper. The trays are of uniform depth and width but of various lengths. The trays should be wide enough to fit into a cabinet drawer in four rows (Fig. 105). Trays about 10 cm (4 in.) wide, for example, form four rows in the size of drawer used for the Canadian National Collection. The advantages of the unit tray system are: it keeps specimens of one species together; it allows the collection to be rearranged or expanded without the necessity of moving every specimen individually; its small trays are more convenient to handle than cabinet drawers; and it restricts damage, because loose specimens can damage only those in the same tray.

Do not overcrowd specimens in boxes, drawers, or trays. Otherwise, they are likely to be damaged if specimens work loose on their pins and swing around, or if you try to pick out an individual specimen.

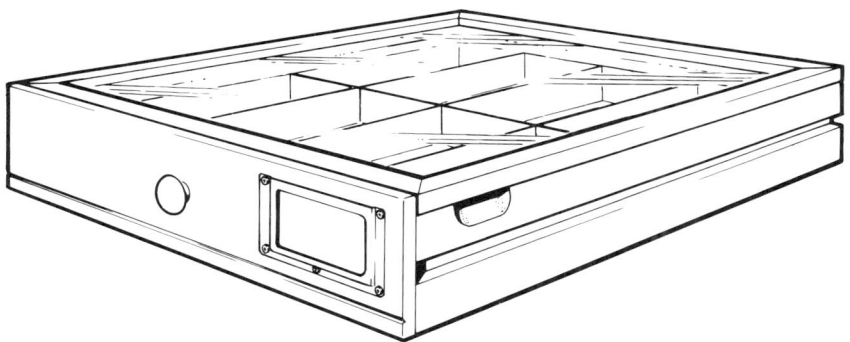

Fig. 105. A glass-topped cabinet drawer with pinning trays.

Despite the use of containers with tight-fitting lids, pests sometimes gain entry to collections. The chief pests in Canada are dermestids, clothes moths, psocids, and ants, but any insects or animals that feed on dried animal material may cause damage. To prevent attacks from these pests put some paradichlorobenzene or naphthalene in each container. Do not put this material loose in the container. If it is in flake form, put it in a small muslin bag and pin it securely in one corner. If the container has a special compartment for such materials, use it. If the material is in solid (ball) form, attach it to a pin by heating the head of the pin and pushing it into the ball, which melts around the pin and solidifies again almost immediately. Renew the naphthalene or paradichlorobenzene in the container as it disappears. If you use too much paradichlorobenzene, it may recrystallize on the specimens and damage them.

A collection that has become infested with pests must be fumigated. Open the boxes or drawers and put them into a box that has a tight-fitting lid. Put paradichlorobenzene at 60 g/m^3 (1 lb/25 cu ft) in the box, close the lid, and leave the box undisturbed for a few days. Alternatively, fill a wide-mouthed vial with absorbent cotton saturated with carbon tetrachloride, ethylene dichloride, or carbon disulfide mixed with carbon tetrachloride, and pin it in a corner of each container, keep the lid closed, and leave the container undisturbed for a few days. Insecticides such as dichlorvos may also be used either in liquid or dry form.

In most parts of Canada, if the containers are stored in a dry place and the insects they contain are fully dried, damage from moisture is not likely to occur. In damp locations or climates, where storage in a dry place is impossible, use silica gel or some other drying agent. Beechwood creosote is effective in preventing development of mold, but it does not prevent the other undesirable results of moisture.

To prevent damage from light and dust, use containers with tight-fitting lids, and if the lids are of glass, store the containers in the dark. There is no cure for specimens that have faded as a result of prolonged exposure to light. Dusty specimens may be cleaned by methods described previously (p. 81).

Storage of a collection preserved in liquid

The chief precaution to be taken in storing a collection of specimens in liquid preservatives is to see that the specimens are always immersed in preservative of the correct strength. The liquid can be prevented from seeping through the stoppers, if they are not made of neoprene rubber, by storing the vials upright. Tight-fitting good-quality neoprene rubber stoppers reduce evaporation, and the vials require filling with additional preservative only after several years of storage.

The best way of storing small vials is to place them upside down in a jar that contains some of the preservative and that has a tight-fitting cap, as described previously (Fig. 95). These jars can be stored in racks a little wider than but similar to the ones shown in Fig. 106 and described below.

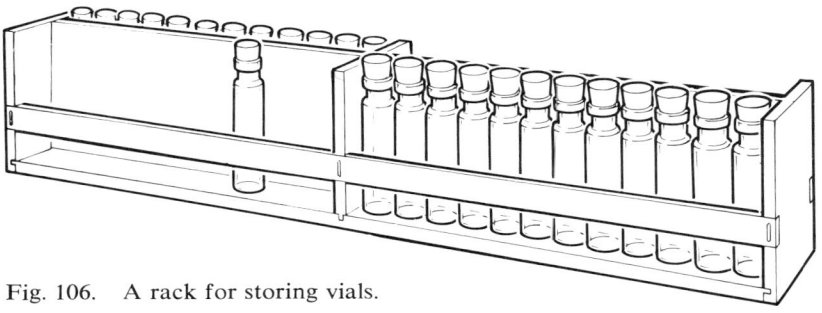

Fig. 106. A rack for storing vials.

Small numbers of vials, or vials that are in regular use, may be stored upright in holes bored in a 2.5 cm (1 in.) board. Each hole should be slightly larger than the vial it is to contain, and the bottoms of the holes are closed by a sheet of card or thin piece of wood glued or tacked to the underside of the board. Alternatively, a shallow box divided into compartments of suitable size by partitions may be used.

Large collections of vials are usually stored in either of two ways. In one method, the vials stand in narrow racks or deep boxes (Fig. 106); each rack is open on the sides, where a stiff wire or thin lath holds the vials in place. Make the ends of the rack higher than the tops of the vials to make handling easier and to prevent the stoppers from accidentally dislodging. Store the racks on shelves of an appropriate size in a cabinet. In the second method, attach the vials to vertical wire screen (Fig. 107).

Storage of microscope slides

Store microscope slides flat, to prevent movement of the specimens or the cover glasses, especially if the mountant is not completely dry and hard. Do not place them directly on top of each other, because they may stick together. Be sure to protect them from dust, light, and dampness.

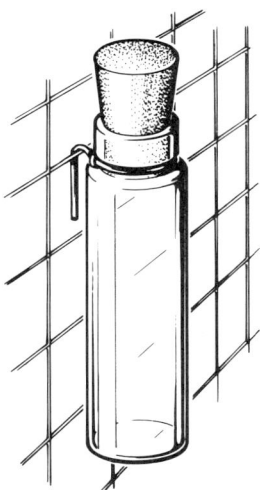

Fig. 107. A screen for storing vials.

The usual slide cabinet (Fig. 108), of wood or metal, has shallow drawers. The drawer should be large enough to hold several rows of slides lying flat. A slide box that has slots in the sides to receive the ends of the slides is shown in Fig. 109. Such boxes must be stored on end, to keep the slides flat. Large collections of slides are usually stored in boxes (Fig. 110) that hold 100 slides.

Fig. 108. A slide cabinet for storing a large collection.

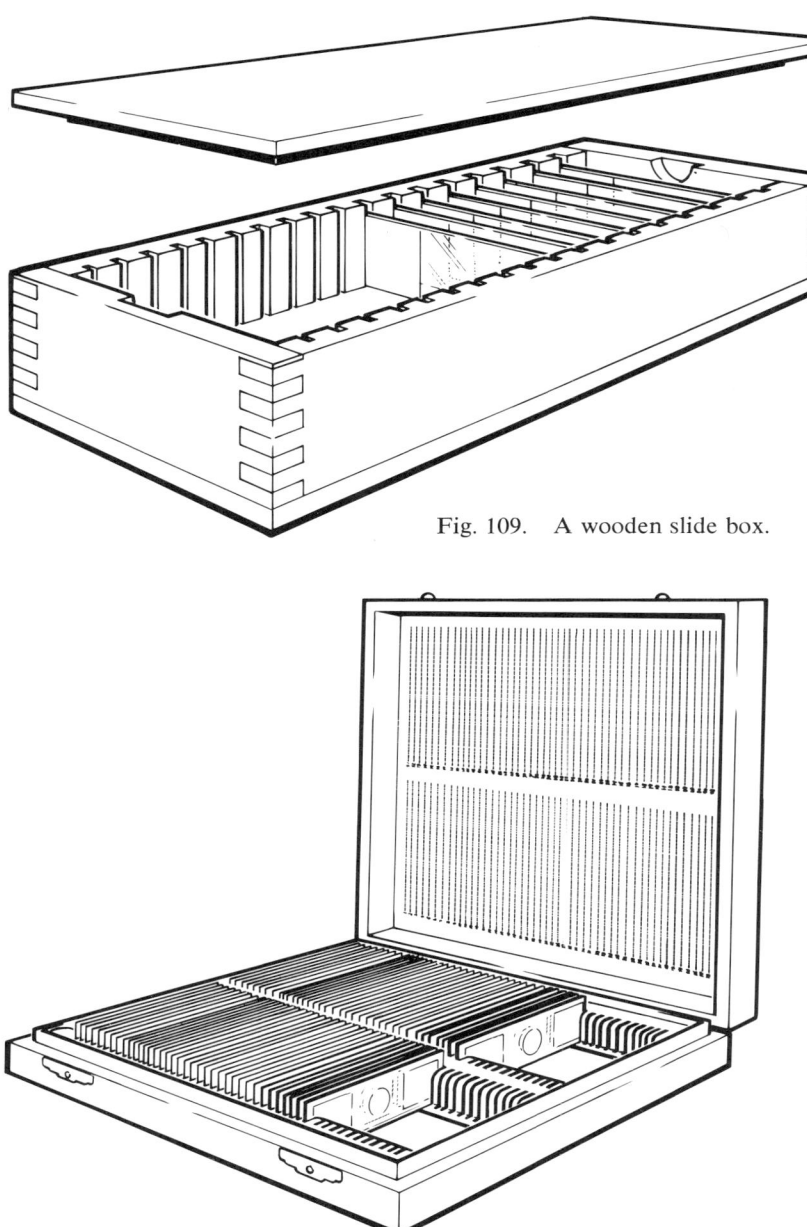

Fig. 109. A wooden slide box.

Fig. 110. A wooden or plastic slide box, used mainly for storage.

Packing insects and arachnids for shipment

Build the outer box in which insects are packed for shipment of cardboard, wood, or metal that is strong enough to survive the normal hazards of transport, such as heavy weights dropping on it. The best way to pack the insects in this box depends on how they are mounted or preserved.

Dried insects in paper envelopes or cylinders layered in pillboxes travel well if packed in a crush-proof box with enough absorbent cotton to prevent them from shifting around.

Insects mounted on pins are often damaged in transit because the sender has not taken the simple precautions necessary to reduce the possibility of such damage. Pin the specimens in a box of suitable size that has a good pinning surface. The quality and thickness of the pinning material are more important than in an ordinary storage box; the material should be thick enough to take and hold pins well. Push the pins in firmly; if the specimens are heavy, push the pins in until they touch the bottom of the box. An additional precaution to prevent pins from working loose is to place a sheet of card the size of the inside of the box on top of the pins and to fill the space between it and the lid with absorbent cotton (Fig. 111). Put a pin at each side of a heavy-bodied specimen or a specimen that has a microvial (Fig. 112) on its pin, to prevent it from swinging around and damaging itself or its neighbor. Spread a thin layer of absorbent cotton on the bottom of the box over the cork to catch parts of insects that break off in transit and to prevent them from damaging others in the box. If you are sending the specimens to a specialist for identification, do not overcrowd the box; leave at least a quarter of the area of the bottom of the box free, for rearranging the specimens and for adding labels. Overcrowding always increases the danger of damage. For shipping, pack the box containing the insects loosely in excelsior, crumpled paper, or a styrofoam material in a strong outer box (Fig. 113). The latter must be large enough to enable the inner box to be surrounded on all sides by a loose layer of the packing at least 8 cm (3 in.) thick. A loosely packed inner box can stand jarring better than one that is packed tightly. Unless the lid of the inner box is very tight-fitting, wrap the box in paper, to prevent particles of the packing material from getting into it. If the specimens are being sent to another country, fix a sheet of cellophane in the box over the insects, so that they can be inspected by customs officials with less possibility of accidental damage.

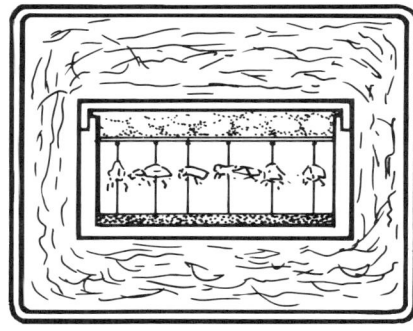

Fig. 111. Packing pinned insects.

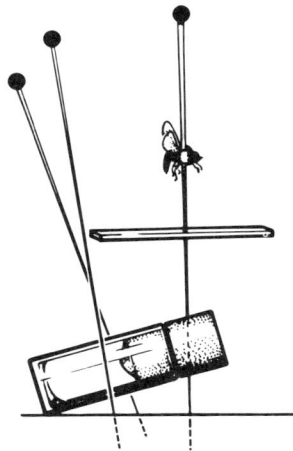

Fig. 112. Bracing a pinned insect.

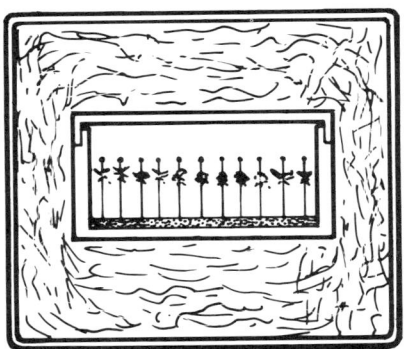

Fig. 113. Packing pinned insects.

Fill vials to the top with the preservative and cork them. If the specimens are delicate, wedge them in the vials with absorbent cotton to prevent them from shaking around. There should be no air bubbles beneath the absorbent cotton, which should not touch the insects. Pack the vials between layers of absorbent cotton in a box packed with excelsior or crumpled paper in a strong outer box, as described previously. Alternatively, plug the vials with absorbent cotton and pack them upright without corks, in a strong jar filled with the preservative and having a tight-fitting cap. Pack the jar in excelsior in the usual manner. Individual vials travel well in hollow wooden blocks or in cardboard mailing tubes, if packed with absorbent cotton to prevent shaking.

Collections (Malaise samples) of insects and other arthropods stored in alcohol may be shipped in polyethylene bags. Drain off the alcohol in which the insects were killed and replace it with fresh 85% ethyl alcohol. Leave the collection in the fresh alcohol for at least 24 hr, or until you are ready to ship it. Drain off the alcohol and immediately transfer the wet collection to a polyethylene bag. Gently squeeze or suck the air out of the bag and tie a tight knot in the neck of the bag to prevent leaking or drying. As a further precaution, place the filled bag (or bags), with some dry paper toweling to absorb any possible seepage, in a larger polyethylene bag that is also tightly closed. Place this bag in a sturdy lightweight container (for example, a mailing tube) with enough packing to prevent movement inside the container, but not to compress the insects. Send the shipment immediately by air, because specimens handled in this way deteriorate after about 2 weeks. As soon as the specimens arrive at their destination transfer them to jars or vials of fresh alcohol.

Boxes specially designed for shipping microscope slides may be obtained from scientific supply companies. The plastic or wooden slide boxes shown in Fig. 109 are available in assorted sizes and capacities, holding 6, 12, 25, 50, or 100 slides. Cover the slides with several soft tissues to hold them firmly yet flexibly in place in the box when the lid is closed. The slides should not rattle audibly when you shake the closed slide box. Tape the lid of the slide box closed for added security. Pack these slide boxes in an outer box with packing. An alternative, and often safer, method is to put each slide between two pieces of wooden molding held together by rubber bands or masking tape (Fig. 114), or to put the slides in single-unit or double-unit slide containers made of heavy cardboard (Fig. 115) and closed with tape. One or several of these wooden or cardboard units can be cushioned in any of a variety of lightweight packing materials in mailing tubes, which come in various sizes and are suitable for shipping by airmail.

It is often necessary to ship living insects from one part of the country to another; this requires special handling. Living insects cannot be shipped into other countries without a special permit. Eggs and pupae can be shipped in tightly stoppered vials or mailing tubes. Pupae should be cushioned in cotton, sphagnum, or other soft material to avoid damage during transit.

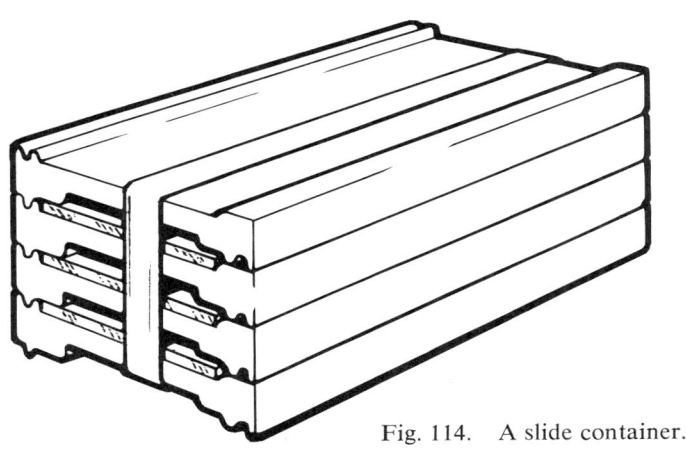

Fig. 114. A slide container.

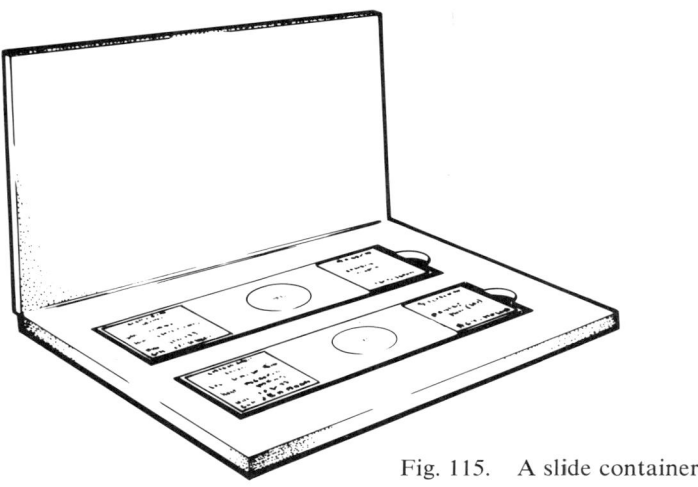

Fig. 115. A slide container.

Larvae should be shipped in plastic bags that are big enough to hold a supply of food plants for the larvae during transit. Place the food plant in the bag first, and then carefully transfer the larvae to the bag. Tie the loose end of the bag securely and put it into a mailing tube or box. Larvae that do not need food during transit can be shipped in damp sphagnum. Use the most efficient means of transportation for shipping living material. Often, you have to make arrangements with the carrier to insure special handling. Freezing, overheating, or dehydration must be avoided, and the package should be labeled with special instructions if they are needed.

Studying the collection

Some equipment not mentioned previously is needed for studying a collection of insects.

Though in the past, most entomological work was done with the aid of a high-powered hand lens, a binocular stereomicroscope is now considered essential equipment for an entomologist. This is a microscope with paired eyepieces and objectives that gives a three-dimensional image that is not inverted. The stereomicroscope differs from the compound microscope, which has binocular eyepieces but single objectives and gives a flat, inverted image at higher magnifications. The magnifications most often used by entomologists are 10–15, 30–40, 60–70, 100–120, and about 200 diam. There are various manufacturers and models of stereomicroscopes, and the selection of one for continuous use is largely a matter of personal preference. Before you buy a stereomicroscope, test various models to select the one most suitable for your type of work. A model with a revolving or drum nosepiece in which the objectives not in use swing forward or upward is perhaps the most convenient; in other models, the objectives sometimes get in the way when you are manipulating or dissecting a specimen. Models capable of various magnifications but with fixed working distances and some with magnification systems are sometimes advantageous, although their resolution drops off quickly above 100×.

The selection of a suitable lamp is important. Many microscope lamps are designed for use with compound microscopes and are not sufficiently powerful or otherwise not fully suitable for use with a stereomicroscope. Be sure to select the type of lamp that is suitable for your use. The lamp should give a strong, uniform illumination at all magnifications.

A compound microscope is needed when magnifications of over 100–200 diam are required, or for studying arthropods that usually are mounted on slides. The design or model of a compound microscope is less important and less a matter of individual preference than that of a stereomicroscope. Binocular eyepieces are helpful if you are using a compound microscope continually. The phase-contrast optical system, and more recently the Nomarski-type interference contrast system are widely used, particularly for the study of weakly sclerotized specimens and of membranous parts.

Excellent drawing attachments, which minimize eye fatigue, and photographic attachments are available for both stereomicroscopes and compound microscopes.

The use of the scanning electron microscope is increasing in studies of insect micromorphology and systematics. This instrument produces a quasi-three-dimensional image within a useful magnification of 50–10,000×. Small arthropods, such as insects and mites, have many minute complex structures that cannot adequately be resolved with stereomicroscopes and they must be studied at higher magnifications. Although these structures can usually be seen with the compound microscope, there are problems of distortion and difficulty of gaining an understanding of the true three-dimensional form. Such structures can be more easily described and interpreted by studying them with the scanning electron microscope, because undistorted material can be examined from different angles. Furthermore, the higher resolving power of the scanning electron microscope often reveals unsuspected detail and previously undescribed characters. Often apparently simple structures are found to be extremely complex.

Besides its value as a tool in the examination and interpretation of small complex structures and fine surface detail, the scanning electron microscope is useful in descriptive taxonomy. The micrographs obtained with the scanning electron microscope are of superb picture quality and have great depth of focus. They can be prepared relatively quickly and are usually more accurate representations than drawings. There is no doubt that scanning electron micrographs will be used increasingly to supplement taxonomic descriptions, especially in groups where the significant features are minute.

Although such sophisticated technology as exemplified by the scanning electron microscope will undoubtedly lead to better and more knowledgeable taxonomy, nonspecialists and amateurs may derive comfort from the knowledge that the key characters of natural groups are usually features that can be seen with the use of a hand lens.

A useful stage for handling pinned insects under a stereomicroscope may be made by gluing two short strips of cork together in the form of an L, or by gluing an ordinary cylindrical cork close to one end of a 6 mm (1/4 in.) cork strip (Fig. 116). Plasticine, molded into a suitable shape and mounted on a 5 × 5 cm (2 × 2 in.) piece of glass or cork, is useful for holding the pin; either the point or the head of the pin may be inserted. The glass or cork provides a flat base.

Fig. 116. A simple stage for handling specimens under a microscope.

Applying the methods

From the collecting and preserving methods already described, the following sections recommend the most effective method for each group of arthropods. Also, some modifications of these methods are given that are useful or advisable for some groups, and additional methods are described that are used only for individual groups.

It is necessary to reiterate that the methods described here are recommended because they are known to be satisfactory and widely used. No one method is necessarily the only or the best one for achieving good results. A collector with any imagination could apply methods recommended for one group, also to another group, particularly methods of preserving soft-bodied arthropods and of making microscopical preparations.

Thysanura

The common species of silverfish inhabit dwellings, but most of the species are in soil, leaf litter, rotten wood, and under stones and bark. Some species inhabit ant and termite nests. These specimens may be picked up with a small brush moistened in alcohol or with an aspirator, or they can be extracted from debris, soil, and leaf litter by using a Berlese funnel (Fig. 36). Kill the specimens in 95% alcohol and store them in alcohol. These fragile insects are easily damaged and should be placed in a shell vial, filled with preservative, and plugged with absorbent cotton. Then they can be stored in larger containers filled with alcohol (p. 20). This is particularly important for species covered with scales, in order to preserve the color pattern.

Diplura

These small, fragile, slender insects are found in similar habitats to Collembola and Protura, and they can be collected by the same methods. Store the specimens in alcohol or mount them directly on slides from 95% alcohol into Faure's or Hoyer's media, and ring them with a sealing compound.

Collembola and Protura

These insects are found in soil, debris, humus, bark fungi, and on foliage. They may be picked up with an aspirator or a moistened brush, or extracted from samples put in a Berlese funnel. Preserve the insects in 70% or 95% alcohol. Collect these specimens in groups of 10 or more specimens of each species, if possible, because some of the specimens are destroyed during the preparations of the slides for study (that is, in some species the eyes have to be depigmented with 10% KOH; in others certain appendages have to be dissected for high-power examination). One of the best methods for examining Collembola is in a temporary mount in 2% KOH.

Collembola are difficult to preserve on permanent slide mounts. A method that gives good, relatively permanent slides consists of warming specimens until they clear in a drop of lactophenol on a slide. A cover slip is then applied and ringed with a rapidly drying sealant after the excess lactophenol has been wiped away with a piece of facial tissue. Other media such as Hoyer's, Faure's, or Berlese's may also be used for mounting, but they are not entirely satisfactory.

Orthoptera and Dermaptera

The true Orthoptera (Saltatoria) occur in a wide variety of habitats, from trees, shrubs, and grasses to swamps and bogs and beneath stones, logs, and debris. Collecting methods vary according to the types being collected: sweeping low vegetation; beating trees and shrubs; and using fingers or forceps to collect specimens under stones or logs. Quick-flying or jumping species may be captured by covering them quickly with a net while they are at rest on the ground (p. 14). When they are covered, move the net handle slightly and they usually hop upward into the net bag, then rapidly sweep and twist the net so that they do not escape. Specialized methods must be used to collect enough specimens of some groups. Collect camel (or cave) crickets by using the bait-jar method: diluted molasses coated on the inside of jars 8–13 cm (3–5 in.) in diam and 13–15 cm (5–6 in.) high. Cover the bottom of the jar with dead leaves and bury the jar to its brim in the ground in rocky or forested areas.

Many species stridulate at night, and the males may be discovered by following their sound to the source. Then, with a flashlight and a net quickly catch the specimen before it flies away. The songs of insects are different for nearly every species. Songs may be recorded, before you capture the specimens, by using a good-quality tape recorder with a microphone attached to a parabola or other device for obtaining the song of an individual.

The habitats where species are collected is often important and should be noted on labels, or recorded in a notebook with a cross-reference number on the data label.

Specimens may be killed with cyanide. Grasshoppers do not die as quickly as some insects do in cyanide bottles. They may appear to be dead, but then later they may revive and damage other specimens by thrashing around. However, specimens should not be left too long in cyanide, because it affects the color, particularly that of green specimens. To kill grasshoppers as quickly as possible, use fresh cyanide bottles. Some specimens, such as camel crickets, tend to shrivel when they dry out and therefore they are best stored in alcohol or preserved by the following process: pass specimens through two changes of 95% alcohol and two changes of absolute alcohol, then transfer them to xylene. Specimens are translucent when they are completely dehydrated, and they can be pinned directly from the xylene. The appendages, although somewhat stiff, can be arranged carefully while the specimens are still wet.

Pin specimens of Orthoptera within 24 hr after their death, and dry them as quickly as possible. Insert the pin through the right side of the pronotum near its posterior edge (Fig. 81), but be careful not to damage the tegmina or wings, particularly in crickets and other singing species, where examination of the stridulatory apparatus on the tegmina may be necessary for positive identification. The long axis of the body should be nearly at right angles to the pin, but with the head of the insect slightly downward. Bend the legs beneath the body to minimize the possibility of breakage and to occupy the minimum amount of storage space. Set the abdomen so that it droops below the wings and is not obscured by the hind legs, because several taxonomic characters are found on the terminal end, and these should not be obscured. Until a specimen is thoroughly dry, support body parts with extra pins so that the specimens stay in the desired position. Very small Orthoptera, such as grouse or pygmy grasshoppers or small crickets, may be mounted on points, or, preferably using minuten pins in *Polyporus*, cork, or balsa blocks, which are then placed on larger pins.

Large (about 3 cm, or 1-1/4 in., or more) specimens should be eviscerated when they are fresh because the internal organs and food in the gut tend to decompose before drying can be completed, and color changes occur. This is particularly important with large green specimens. The most satisfactory method is to break the membrane dorsally between the head and thorax, remove the internal contents with fine curved forceps, swab out the inside with cotton, then insert a mixture of one part boric acid to three parts talc. Shake the specimen to distribute the talc. Use a drop of adhesive to hold the head close to the thorax. This method promotes quick internal drying and preserves color better than other known methods.

Evisceration can also be done by making an incision through the first three abdominal sterna along the midline with fine, sharp scissors and removing the thoracic contents of the abdomen with fine straight forceps. For very large specimens, the body cavity can be packed lightly with cotton.

If specimens cannot be pinned soon after they are killed, eviscerate them, and place them in layers in Cellucotton in small, strong boxes. Later they can be relaxed, pinned, placed in the desired position, and allowed to dry. One important thing to remember is never to allow the specimens and the data concerning them to become separated.

Dictyoptera (cockroaches and mantids) should also be pinned in the same manner as Orthoptera. Cockroaches are usually found in buildings; they may be collected in bait jars by the same method as used for camel crickets. Mantids are usually found in late summer on vegetation. Green females should be eviscerated after you have slit their abdominal sterna.

Phasmatodea (stick insects) are found on trees and are usually taken by beating the trees. Leave them unpinned and stored in cellophane envelopes with heavy cards to prevent breakage. Write all pertinent data on the card. If you pin your specimens immediately, arrange the appendages and support them with a stiff card (Fig. 117) until they are dry. Remove this support before labeling.

Dermaptera are often found under sticks, stones, or other debris. Pin them in the same manner as Orthoptera.

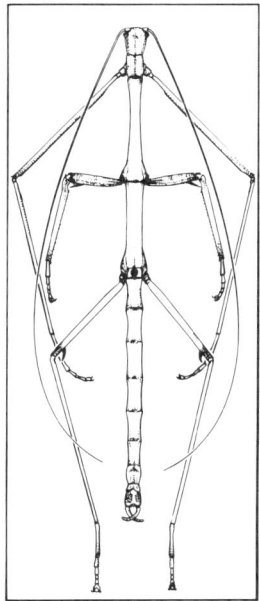

Fig. 117. A specimen (Phasmid) supported on a card for drying.

Plecoptera

Stoneflies occur near the streams, rivers, and lakes in which their early stages develop. Adults of the fall, winter, and spring species may be collected on bridges, tree trunks, or other objects, and on snow and ice near streams and rivers. Some summer species are attracted to lights, and others may be collected by beating or sweeping the vegetation near the shores and banks. The nymphs are found under stones and debris, usually in well-aerated water.

Preserve adults and nymphs in alcohol. For study, snip the genitalia from the abdomen and clear them in a 10% solution of caustic potash (KOH). Keep the dissected parts in the vial with the specimen. Slide-mounted genitalia are difficult to examine except in the plane in which they are oriented.

Isoptera

The four species of termites that are found in Canada occur only in the southern parts of Ontario and British Columbia. Examples of the different castes of a species should be collected from the nest and preserved in alcohol.

Corrodentia

Species of psocids are found on vegetation, on and under bark, and in debris on the ground. Some species infest dwellings, buildings, and granaries, where they are commonly found in undisturbed books and paper, straw or similar material, and cereals or cereal products that contain starch, on which they feed. They may be collected by beating, sweeping, searching, or by drying debris in a Berlese funnel. Specimens are usually preserved in alcohol, although very large species are sometimes pointed on pins in order to preserve some of the color pattern. Mount smaller specimens on microscope slides and clear them by the same method as recommended for aphids.

Mallophaga

Most species of biting lice are ectoparasites of birds. However, several species occur on mammals and these may be collected by the same methods as described for Anoplura. The species found on birds are extremely diverse; the various species usually occupy separate microhabitats (for example, the back of the neck for one species, the wings for another species) on the same host. When you have collected the birds, place them in paper, cloth, or polyethylene bags, but only put one species of bird in a bag, because biting lice are very host specific. When the host dies, the lice usually move to the tips of the feathers, where they can be easily collected. However, you should search carefully with forceps the entire surface of the bird and its feathers in order to obtain all specimens, especially of the relatively sessile species (*see* the section "Collecting Ectoparasites of Vertebrates," p. 67).

Mallophaga may be stored temporarily in alcohol, but they should ultimately be mounted on microscope slides in Canada balsam. Use the standard KOH–balsam technique described for Anoplura or the lactophenol process to prepare whole slide mounts. If you use the latter method, warm the specimens in distilled water until the alcohol has been replaced by water and the abdomen becomes distended. Then transfer the specimens to lactophenol and boil them vigorously in a fume hood until they clear. Transfer the cleared specimens to a dehydrating and hardening solution that is miscible with Canada balsam. This solution consists of 40% glacial acetic acid, 40% oil of cedar, 10% oil of wintergreen, and 10% oil of lilac. Slit the integument of the abdomen with a sharp fine needle while the specimens are in this solution, and gently press out the liquefied contents of the abdomen with a blunt needle. Be careful not to break the setae. Stain very pale or weakly sclerotized specimens with acid Fuchsine; always leave some specimens of each species unstained. Add a few drops of acid Fuchsine to the acetic acid – oil solution. If overstaining occurs, wash the specimens in a clear solution of the acetic acid – oil preparation until the desired density is reached. The time required for dehydrating and hardening varies with the size of the specimens, but best results are obtained by leaving specimens in this solution for about 12 hr. Transfer the specimens directly from the solution to a drop of Canada balsam on a microscope slide. Arrange appendages and position the specimen before the cover slip is added. Dissect the male genitalia of at least one specimen in acetic acid – oil and mount them under a separate cover slip. If the specimens become milky or opaque when placed in the Canada balsam, they are not completely dehydrated. Dry the slides in an oven until they are firm.

Anoplura

The sucking lice constitute a small order, and are exclusively ectoparasites of mammals. Some of the lice of small mammals, such as rodents, are minute and hard to find. They often remain attached to and die on the host; in fact, many valuable collections have been made from skins preserved in museums. The usual method of collecting involves a meticulous search of an animal's fur, with fine forceps. Sometimes this is best done under a dissecting microscope. If you discover nits (eggs) cemented to hairs, search again for lice. Even after the most diligent searching many lice may be missed, and the Hopkins technique may prove useful. This involves rough skinning the mammal, drying the skin, and storing it for future examination. When you want to remove the lice, cut the skin into small pieces and soak them in a 5% solution of sodium hydroxide until they become soft enough that the hair can be removed with a blunt knife. This takes about 15 min or a little longer. Place the partly dissolved hair in a beaker, add about a quarter of its bulk of 5% sodium hydroxide, and place the beaker in a pan of water and boil it until the hair is completely dissolved—a process taking about 30 min. Filter the contents through very fine mesh stainless steel gauze. Wash the residue in the gauze with a fine stream of water to remove the smaller particles, and wash the remainder into a watch glass or petri dish and examine it under a low-power microscope (or, if you are in the field, use a hand lens). Pick out the parasites, and preserve them in alcohol. To make temporary microscope slide mounts, soak the specimens for 24–48 hr in liquid phenol (phenol crystals dissolved in a small amount of absolute alcohol). To prepare specimens as permanent mounts, follow the same methods as described previously for Mallophaga, or clear the specimen in 10% KOH for about 24 hr. Slit the integument with a sharp needle and, after the specimen has cleared, gently press out the liquefied contents of the abdomen with a blunt needle. Then wash the specimens in water, dehydrate them through 70%, 95%, and absolute alcohol, and immerse them briefly in cedar oil or xylol before mounting them in Canada balsam or a similar material. Very pale specimens may have to be stained lightly by adding about six drops of acid fuchsin to distilled water in a cavity slide for 1 hr. Some specimens of each species of Anoplura should always be left unstained, and the male genitalia of at least one specimen should be dissected.

Ephemeroptera

In their immature stages mayflies are aquatic, and the nymphs may be found in various freshwater situations. Nearly 600 species have been found in North America. In lakes and ponds, the nymphs crawl freely on the bottom on submerged debris and stones, or they may burrow in the mud. In streams, they are more likely to be found on the undersides of stones, pieces of wood, or other matter, often in swiftly flowing water. Nymphs may be collected by hand, or by dredging or raking the bottom of the pond or stream. Place mature nymphs (which usually have conspicuous blackish wing pads) directly in alcohol, unless you want to rear them to the adult stage. In the latter event, confine the specimens in cages placed partly submerged in the water, or in containers in which the natural environmental conditions (temperature, aeration) are as closely approximated as possible.

Adult mayflies may be collected in many locations, but specimens are usually most abundant near the waters from which they have emerged. Many species appear in swarms along the banks of streams during the early hours of the morning or toward evening. Others swarm in bright sunlight, but swarming at any time takes place usually only in the absence of wind. When the adults are not swarming, they rest motionless on the underside of foliage and in sheltered situations. In such places you may find many species that do not swarm. Sweeping or beating foliage may disclose the presence of specimens, especially if they are not numerous enough to be seen easily. Specimens collected in these ways are likely to be damaged unless you are very careful. Avoid these collecting methods whenever possible. You can obtain more satisfactory results by collecting the specimens individually by netting them from swarms, or by collecting them at light, to which most of the adults are strongly attracted. In general, light traps are not satisfactory and few specimens from such sources are found in satisfactory condition for preservation. When collecting mayflies, be careful to distinguish between the fully developed adults and the subimagoes (which are usually recognizable by their opaque wings and rather short appendages). If subimagoes are taken, keep them alive in a paper-lined jar until their final molt has taken place and the adult has reached full development of sclerotization and pigmentation (usually several hours). A handy device for collecting subimagoes, or adults that are slightly immature, is shown in Fig. 118. With a little practice you will discover that you can trap the resting mayfly by slipping a jar quickly over the specimen; the funnel cap prevents the captured specimens from escaping while others are still being collected. A collecting kit should include several jars with ordinary screw caps and one or two with the interchangeable funnel tops.

Fig. 118. A jar for collecting specimens of Ephemeroptera.

Adult mayflies may be preserved either in alcohol or by pinning. Because the specimens are very fragile, both methods must be very carefully performed to achieve satisfactory results. If you preserve them in alcohol, place only a few specimens in each vial, and handle them very carefully, especially in transferring specimens from one container to another, because legs and other appendages may be lost. Pin only freshly killed, fully matured specimens and handle them with special care. So treated, pinned material is satisfactory for study, and is preferred by some workers. In studying mayflies it is advantageous to have both pinned and alcohol-preserved material. Therefore, the general collector should try to secure series of each species and preserve several specimens by each method. Some collectors put adult mayflies in small cellophane envelopes, but this procedure is not generally considered satisfactory.

Odonata

Over 400 species of dragonflies are found in North America. An ordinary aerial net is inadequate for extensive dragonfly collecting, particularly for the strongly flying species. A net with a large ring, 46 cm (18 in.) or more in diam, a handle 1 m (3 ft) or more long, and a bag of wide mesh is more convenient. When you swing the net at a flying dragonfly, sweep from behind the insect; otherwise, the insect may dodge the net. Dragonflies have keen sight and are readily alarmed by rapid movements; therefore, make slow and cautious movements until the actual movement of swinging the net. Certain species have definite patrol routes and others have favorite perches—facts that can be utilized by an observant collector to his advantage.

Most species fly only in the sunshine, though there are a few that are active at dusk or even after dark. The neighborhood of marshy ponds is the most productive. Streams, especially those with gentle rapids alternating with pools, have a fauna of their own, as have sphagnum ponds, particularly those with floating margins, spring runs, spring-fed pools, and oxbow ponds. The rarer species are likely to be found in the more unusual kinds of habitats, and the most unlikely looking places should not be neglected.

Specimens may be killed in cyanide, but do not leave them too long in its fumes because of color changes. If possible, specimens may be kept alive for a day or two after they have been captured so that they empty their intestinal tracts, which helps retain colors better in some specimens.

There are two common methods of preserving dragonflies. One is to put them in vials of 95% alcohol. This method has the advantage of preserving the colors almost perfectly. The other is to store the specimens dry, either in paper or cellophane envelopes or pinned and spread. Dried specimens often become badly discolored and brittle. However, colors are fairly well preserved if the specimens are kept in a warm, dry place. Because spread dragonflies take up so much space in a collection, it may be advisable to pin and spread only a few specimens of each species, leaving the rest in envelopes. Colors can be better preserved in papered specimens by rapid drying in some type of improvised desiccator. A pinned specimen has the pin inserted in the midline of the thorax between the bases of the hind wings.

Preserve the nymphs in alcohol in vials. Exuviae, or cast-off skins, particularly of Anisoptera, are valuable taxonomically. Glue each one to a narrow strip of card, and pin it in the collection. Cast-off skins of Zygoptera (damselflies) should be preserved in alcohol.

Thysanoptera

Thrips are found in all kinds of growing and dying vegetation: in flowers, soil, and mosses and on foliage, fruits, twigs, bark, and decaying wood and fungi. Many species are confined to one kind of plant, or to a few related kinds. Consequently, host data are important. Collect the insects from one kind of plant or habitat at a time and label them accordingly. If the plant cannot be identified in the field, preserve it in a plant press and it can be identified later. A newspaper tied tightly between two pieces of stiff cardboard is a satisfactory plant press. In such cases, carefully cross-label both the plants and the insects. One method of collecting thrips from vegetation is to place a sample of one plant species in a tightly closed container. A good-quality paper bag closed with several paper clips is satisfactory, but a glass jar with a tight-fitting stopper or screw lid is better. Do not mix samples or plant species; put each one in a separate container.

There are several methods of extracting thrips from samples of vegetation, soil, rotting wood, or other materials inhabited by these insects. The best method is to use a Berlese funnel (Fig. 36); this piece of equipment is almost essential. Put one pure sample in the funnel at a time, and clean the funnel thoroughly before it is used for another sample.

If Berlese funnels are unavailable, carefully empty the sample of vegetation onto a large sheet of white paper under a strong light. By agitating the vegetation from time to time you encourage the thrips and other insects to crawl out on the paper, so that they can be picked up with forceps or a moist brush. If you use glass jars for your samples, many specimens can be extracted by holding your hand over the open end of the jar and by picking up the thrips from your palm. In the field, thrips can be collected by striking a flower or piece of foliage on the palm of your hand or on a sheet of white paper or by beating limbs of trees or shrubs with a stick over a large white cloth or a beating tray. Samples of soil, rotting wood, and fungi can be slowly strained through a sieve onto a white cloth or paper.

The most common preservative used for thrips is AGA (*see* "Formulas," p.169). However, the specimens tend to become distended and discolored and, consequently, they should be mounted within a month of when they were collected. Alcohol is a good preservative, but specimens preserved in it tend to become brittle and difficult to mount. Drop the specimens into the preservative alive and store them in vials.

Specimens can also be preserved and stored in lactophenol solution, which clears the specimens well for microscopic studies. Through the years many methods of preparing slides of thrips have been suggested. However, probably the simplest and most effective methods are those used for the Aphidoidea, with the use of Hoyer's medium or Canada balsam as a mountant. Black thrips should be bleached for 24 hr in 2% KOH before applying the preceding methods. Slides produced by either of these methods are satisfactory for research and identification, provided the appendages and wings have been carefully spread with a very fine dissecting needle before the cover slip is applied. Use only the correct amount of medium: the medium should not be thicker than the specimen. Another simple, but fairly effective, method is to immerse freshly collected specimens in oil of cloves for 24 hr. Then the specimens can be mounted in Canada balsam, which has been thinned to a manageable viscosity with terpineol (oil of lilac). This latter method is useful for collecting large numbers of thrips, as might occur on an ecological project.

Hemiptera (including Heteroptera and Homoptera)

The species included in this group are the true bugs, leafhoppers, cicadas, aphids, and their allies. Most of these species are free-living on plants, some are found under loose bark, among the matted vegetation at the base of grasses and other plants, or in vegetable debris, and a few are parasites of man and birds. The best months for collecting the adults are June, July, and August; earlier and later in the year, the species are immature. Sweeping and beating are productive methods of collecting. Host-plant data are essential in order to identify some groups. This can be obtained by searching the host and handpicking the insects with an aspirator. Specimens from debris are easier to find by putting samples of debris in a separator. Many species are attracted by artificial lights at night, and these may be captured by traps or picked up by hand.

The first method described applies to preserving and mounting Hemiptera other than Aleyrodoidea, Aphidoidea (aphids), and Coccoidea (scales); and the second method applies to these last three groups.

Preserving and mounting Hemiptera Cyanide and ethyl acetate are satisfactory killing agents for the species in this group. Kill aquatic and semiaquatic species and store them in alcohol until they can be mounted. Semiaquatic species, because of their dry condition, may be placed directly into killing bottles and temporarily stored in pillboxes. Do not use alcohol to kill or store terrestrial Hemiptera, particularly the Miridae that are extremely delicate, because it deforms, bleaches, and dehydrates them. But, use alcohol to preserve all immature forms. For temporary storage, adults, except those of aquatic species, should be stored in layers in pillboxes.

Point or pin adults directly. A general rule to follow is to point all specimens 13 mm (1/2 in.) long or less unless they are very large and robust (that is, 6 mm, or 1/4 in., wide or more), and point specimens more than 13 mm (1/2 in.) long if they are extremely slender (too slender to take pins without distortion). If this rule is followed, most of the species of Hemiptera found in Canada would be pointed.

Arrange protruding legs and antennae that might get broken off during mounting near, but not against, the body. In Cicadidae and Psyllidae it is important to have a clear side view of the terminal abdominal segments. Consequently, arrange the wings on the left side of the body to point upward from the body in Psyllidae, and to be spread in Cicadidae.

Do not use pins smaller than No. 1 for direct pinning. Insert the pin in the body in different locations for different groups, because of the diversity of structure of species. Insert the pin through the scutellum near its anterior margin just to the right of the midline in species in which the scutellum is large and not covered by other structures (that is, in specimens in which the scutellum is at least twice as long and twice as wide as the diameter of the pin (Fig. 80)). For species in which the scutellum is small or mostly covered by the enlarged pronotum (for example, Notonectidae), insert the pin through the pronotum near its posterior margin and just to the right of the midline.

Point most Membracidae, which have an enormously enlarged pronotum; point all Cicadellidae; other large specimens that may be pinned directly should have the pin inserted just to the right of the midline about a third of the length of the body from the front (Fig. 79). Affix a point to the underside of the right side of the thorax. Flat specimens do not need the tip of the point bent down, but with other specimens it is important to bend the tip to the correct angle of the thorax. Affix small Cicadellidae to the straight tip of the point on the mesothoracic coxae. For larger Cicadellidae straighten the right hind leg and affix the bent point to the right side of the thorax.

Preserving and mounting Aphidoidea, Coccoidea, and Aleyrodoidea Collect insects of the groups Aphidoidea, Coccoidea, and Aleyrodoidea by searching and handpicking them from the infested host plants. Collecting by other methods does not produce the host data that are often needed for identifying the specimens, although sweeping or beating is sometimes useful to reveal the presence of insects on a host plant. Notes on the natural colors of the living insects are helpful for making identifications and they are also useful for future recognition in the field.

Aphidoidea (Aphididae, Adelgidae, and Phylloxeridae): Colonies of these insects may occur anywhere on the host plant. It is often convenient to remove the infested part of the plant and to preserve it in alcohol with the aphids. By doing this you are more likely to collect all the stages present in the clone. Another method of collecting these insects is to place the infested part of the plant in a plastic bag and to store it in a cool place in the laboratory for a few days before preserving it. This method often yields additional winged morphs, or late in the season it may produce sexuales; sometimes parasites and hyperparasites emerge. A useful preservative and clearing agent for these insects is lactophenol. This solution, which is prepared by dissolving 2.25 kg (5 lb) of clear phenol crystals in 1 litre (1 qt) of lactic acid, is an excellent clearing agent for aphids, adelgids, and phylloxerids, and it removes all traces of noncuticular internal parts. Aphids have been stored in this material for 7 years without apparent deterioration of the cuticle. If you use lactophenol for storage, label the outside of the vial.

Before specimens of Aphidoidea can be identified, they must be cleared and mounted on microscope slides. Specimens that have been preserved in alcohol should be warmed in distilled water until the alcohol has been replaced with water and the abdomens have become distended. Then decant the water, add lactophenol, and boil the specimens vigorously in a fume chamber. When the specimens have been cleared, mount them either in Hoyer's medium or in a neutral solution of Canada balsam in xylene. If you use Hoyer's medium, or a similar medium containing chloral hydrate and gum arabic, the specimens can be transferred directly from the lactophenol clearing agent to a drop of medium on a slide. Place the specimen on the slide with the dorsal side up and carefully spread the appendages with the beak extended between the coxae. Add a cover slip and dry the slide in an oven until the medium is firm. Slides made with any medium containing chloral hydrate and gum arabic should be ringed with rapidly drying sealant. Several satisfactory sealants are available from most biological supply houses. Unless the mounts are carefully ringed, air creeps inward from the edge of the cover slip and destroys the specimen for microscopical study.

Canada balsam is the best mountant for preparing permanent slides such as the ones used for museum specimens. When you use Canada balsam, decant off most of the lactophenol, and add a dehydrating and hardening solution that is miscible with the balsam. This solution consists of 40% glacial acetic acid, 40% oil of cedar, 10% oil of wintergreen, and 10% oil of lilac. The duration of treatment in this solution varies with the size of the specimens, but the best results are obtained usually by leaving the specimens in this solution for about 12 hr. Transfer the specimens directly from the solution of acetic acid and essential oils to a drop of balsam, carefully spreading the appendages before adding the cover slip. The amount of mountant to use depends on the size of the specimens, but the medium should not be thicker than the head of the specimen. If you use Canada balsam as the mountant, thin it to a manageable viscosity with oil of lilac. If the specimens become milky or opaque when they are in the balsam, dehydration is not complete. Dry the slides in an oven until they are firm. Staining is unnecessary, especially if you are using a microscope equipped with attachments for phase and interference microscopy. Do not remove embryos before you mount aphids, because embryonic structures are often useful for identification and taxonomic research.

Coccoidea: Usually it is possible to identify only the adult females. They are either sessile or motile, and may be found on nearly every part of the host plants. The best way to collect and preserve sessile species is to remove the infested parts of the plant and to leave them to dry in a receptacle that is not airtight. However, the motile species can be collected directly into alcohol or a lactophenol solution. You must make microscope slides in order to identify these species. Follow the methods given for the Aphidoidea. To clear the dried specimens, boil them in lactophenol in a fume chamber. Staining is generally unnecessary if you have a microscope equipped with phase and interference attachments for examining your specimens. Male scales, unlike those of females, usually can be preserved satisfactorily in alcohol.

Aleyrodoidea: Whiteflies, unlike other Homoptera, have a pupal stage in their life cycle. This is the stage mainly used for their identification. To collect these pupae, place leaves infested with pupae in ventilated containers. Preserve the adults that emerge in alcohol or in lactophenol. You must make microscope slides in order to identify these species. Follow the methods described for the Aphidoidea. Clear the dried pupae by boiling them in lactophenol in a fume chamber before you make the slides.

Megaloptera

To collect dobsonflies and alderflies sweep or beat the vegetation near the aquatic habitats of the larvae. Most species are attracted to light. However, because they are weak fliers, set the light up near water. Kill the specimens in cyanide or another killing agent and preserve them in alcohol. It is always wise to pin and spread a few specimens of each species. Dobsonflies can give you a sharp bite, so be careful when you handle living specimens. Preserve the immature stages in alcohol.

Neuroptera

Lacewings, antlions, and their allies are usually collected by netting, sweeping, and searching; occasionally some species come to a light. Kill the specimens in cyanide or another killing agent, pin them, and if necessary spread them. Because the male genitalia are used as a means of identification, preserve part of the series in alcohol.

To prepare the male genitalia for study, remove the abdomen from the specimen and clear it in a 10% solution of KOH for about 24 hr. Then transfer it to alcohol and gently flush out the liquefied contents of the abdomen with a jet from a hypodermic syringe. Use a fine syringe needle (27 gauge) for this operation. When the abdomen is completely cleaned out, use the same method to evert the structures. Stain the abdomen in a drop of 5% chlorazol black E aqueous solution for 2 or 3 min and return it to the alcohol to flush it out again. For examination, place the genitalia in a drop of glycerin on a microscope slide. To avoid flattening or distorting the structure, support the cover slip by placing a few pieces of a broken cover slip under its edge. Permanently preserve the genitalia in a microvial (*see* "Microscopical Preparations," p. 102).

Mecoptera

Scorpionflies are found mainly in damp, wooded places; some wingless forms may be found on the snow. They fly mainly during the day, though some will come to a light after dark. Use a net to collect the winged forms from low foliage in shaded areas, or use a Malaise trap in a suitable habitat. You can pick up the small wingless species (*Boreus* spp.) with forceps or you can catch some in pan traps (p. 30). Kill all forms in cyanide or another killing agent. Before you put a specimen in a killing bottle, hold it in your hand for a minute until it emits a brown liquid; otherwise, this liquid will be released in the bottle and it will foul the bottle and the other specimens. Pin scorpionflies through the middle of the thorax and spread at least part of the series. The small wingless species may be pointed. Preserve the immature stages in alcohol or another liquid preservative. As in the preceding group (Neuroptera), the male genitalia are used as a means of identification, and the same procedure can be followed for preparing and preserving them.

Trichoptera

Caddisflies are found mainly near lakes, rivers, and streams, though some species fly long distances from their breeding grounds. Over 1000 species occur in North America. They may be netted as they fly in the evening or after dark or swept by day from vegetation near the water. They are strongly attracted by light, which is the most productive collecting method.

Preserve specimens in alcohol or pin them. It is best to pin a few specimens of a species through the thorax while they are fresh, and to preserve a larger number in alcohol. The pinned specimens may be spread, though this is not essential. If you do not spread them, bend the wings (Fig. 119) downward so that the genitalia can be readily examined. Always pin species of the genus *Leptocella* (white insects with yellowish or grayish transverse bars on their wings), because they lose their pattern in alcohol.

Rearing caddisflies from larvae is easy and profitable; rare species may be obtained in this way. Pupae are valuable because larval skins are associated with them, and the genitalia of advanced pupae are identifiable. In this way larvae and adults can be associated. Preserve the pupae in alcohol.

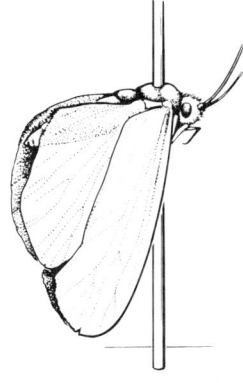

Fig. 119. An unspread specimen (Trichoptera) showing position of wings and abdomen.

Lepidoptera

Though the Lepidoptera of some parts of Canada are fully known, systematic collecting provides valuable information on the ranges and habitats of the species, and may turn up species that have not yet been discovered in Canada. Microlepidoptera are inadequately known in all parts of Canada and can profitably be collected anywhere. The life histories of most Canadian Lepidoptera are very inadequately known, and the assembling of a collection of well-preserved, identified specimens of eggs, larvae, and pupae, and information on their food requirements and preferences are among the most useful and also the simplest contributions that can be made by an individual collector.

In spite of their ability to fly, Lepidoptera are often closely tied to specific types of habitat. Most species of Lepidoptera have short flight seasons, and many can be taken in good condition for only a week or two during the year. Thorough collecting in any locality therefore requires repeated searching of each type of habitat at all suitable seasons.

There are four principal methods of collecting adult Lepidoptera: with a net, by attracting them to artificial light, by using baits or natural attractants, and by rearing from the early stages. Each has advantages and disadvantages.

Nets are most often used for day-flying Lepidoptera, but they are also used in conjunction with a flashlight or lantern to catch nocturnal species, particularly those attracted to flowers. Many diurnal Lepidoptera are attracted to flowers, some to wet places on roads; a few butterflies can be attracted by the use of a dead specimen of the same species as a decoy. Early morning collecting is usually unproductive; from late morning until midafternoon or later is the best time for butterflies. Many species of moths fly most freely just after sundown. Warm, sunny, calm weather is usually the best, though diurnal moths are less strongly affected than butterflies by the absence of sunlight.

Collecting at a light is probably the most productive method of obtaining moths. For best results use a strong light placed so that it lights up a wide area, and place a sheet or other reflecting surface behind or below it. Light traps catch Lepidoptera in large numbers but unless the killing or stupefying agent is strong enough to act quickly the condition of the specimens is poor. Collecting at a light should not be confined to the early evening: some species rarely appear before midnight, and there is usually a definite change in the species collected as the night progresses.

The most widely used bait for Lepidoptera is "sugar." You can vary the recipes according to your preferences and the materials you have with you. "Sugar" is attractive chiefly to Phalaenidae (= Noctuidae) but a few species of other families also visit it. The "sugar" is a sweet-smelling liquid made up largely of sugar in some form that has been partly fermented. There is no standard formula; most collectors develop their own. The mixture usually consists largely of brown sugar, molasses, or both, to which you add stale beer. You may want to add mashed bananas, mashed overripe peaches, fermenting fruit juices, and a small quantity of rum, asafetida, or an ester such as amyl acetate. The mixture becomes more attractive if left overnight or longer to ferment; fermentation is assisted by adding a small quantity of yeast. Important features of the mixture are that it must be aromatic, giving off odors of lower alcohols and esters, which are fermentation products, and it should be thick enough not to run off or soak into the trunks of the trees to which it is applied. Apply the mixture with a brush to tree trunks, fence posts, and other objects at dusk, and visit these locations at intervals after dark, when the moths and other insects can be found imbibing. Paint a patch of the mixture about the size of your hand on each tree trunk at a convenient height, preferably on the side away from the wind. For convenience, select trees that form a closed circuit. Such a "bait line" grows much more attractive if you use it consecutively night after night over a long period. If you use it in the same place year after year, additional species continually appear. The selection of a good locality for "sugaring" is important. The characteristics of such a locality vary from region to region and have to be determined by experience. In general, choose a locality that is sheltered from the wind and where the vegetation is not too dense, preferably mixed growth. In a wood, for example, the best situations are the margins, paths, and glades. The numbers of insects attracted vary greatly, depending on the locality, season, weather, and presence or absence of natural baits. Usually, nights that are light are good for "sugaring." Insects feeding at a "sugar" patch often become sluggish, probably because of the effects of the alcohols. The moths usually fly outward and downward when disturbed, or they simply drop to the ground. The method of catching a specimen is to hold an open killing bottle or a net immediately below it and, if necessary, to move the bottle or net up slowly until it touches the insect. "Sugar" is, of course, only a substitute for such natural attractants as sap, nectar, honeydew, and decaying fruit, and often these can be used to advantage. Birch sap is a particularly effective attractant in the spring, as is willow sap on the prairies. The flowers of willow, lilac, sweet rocket, evening-primrose, and milkweed deserve special mention. Deep bell-shaped flowers with a heavy perfume are especially attractive to Sphingidae.

Males of some species of moths gather around unfertilized females, and can conveniently be captured. In Britain, this method of collecting is called assembling. An unfertilized female may be used instead of other bait in a bait trap.

Besides the previously mentioned methods of collecting, some other activities are sometimes quite productive. Moths can often be found resting during the day, especially on tree trunks; they may also be found near electric lights that attracted them the night before. Hepialids can be found at appropriate seasons and times of day forming mating swarms near their food plants. Sweeping vegetation with a net is a good way of collecting larvae, but it is useless for collecting adults, because specimens obtained in this way are invariably badly rubbed. A serious collector always carries a killing tube with him, just in case he sees a specimen in some unlikely place at an unusual time.

Rearing Lepidoptera from the early stages has several advantages. By this means, specimens can be obtained in perfectly fresh condition, an important feature in this order; females can be obtained in as large numbers as males; a series of a species can be obtained from eggs laid by one female; and, most important, the life histories and the food-plant preferences of the species can be learned. Early stages for rearing can be obtained by inducing a captured female to lay eggs or by collecting eggs, larvae, or pupae in their natural surroundings. The first method has the advantage of yielding many unparasitized early stages of known identity: a long series of adult specimens can be obtained, and specimens of eggs, larvae, and pupae can be preserved, without risk of being mistaken for similar species and without the need for success in rearing some individuals to the adult stage for identification. Female Lepidoptera caught in nature will usually lay fertile eggs; reared females must be mated, either with reared males or with wild males attracted from the surrounding country. The second method may shorten the rearing process, reveal unknown food plants of the species, and is likely to yield parasites that, if preserved, are of value to the hymenopterists and dipterists.

Most species of moths lay eggs readily in confined quarters. Fairly high humidity, darkness, and the presence of liquid food and the larval food plant are all factors favoring oviposition. Most butterflies lay eggs only in sunlight, and many species need enough room for flight before they lay; many oviposit only on their food plant. The care of larvae is simple: clean surroundings and an abundance of fresh food. Handle the larvae and pupae as little as possible.

Lepidoptera emerging from the pupae need plenty of room, and should not be interfered with until their wings have expanded and hardened.

Use cyanide to kill Lepidoptera. Because these insects are so easily damaged, take certain simple precautions: keep the bottle clean; do not put other insects in with Lepidoptera; place some absorbent paper or absorbent cotton in the bottom of the jar; put in only a few specimens at one time; and never carry a bottle with specimens lying loosely in it any distance (pack the specimens between layers of absorbent cotton if you have to carry them far). Liquid killing agents are reasonably satisfactory if cyanide is not available. But use them carefully because they tend to make specimens stiff and hard to spread, they may wet the specimens, and an underdose of fumes may stun the insect without killing it so that it may damage itself or escape.

Adult Lepidoptera should be pinned and spread as soon as they have been killed. When this is impossible, they may be stored temporarily in triangular paper envelopes, as described previously. In an emergency, specimens can be layered on a sheet of absorbent cotton and covered with a layer of facial tissue in a fairly large, shallow box. It is better to pin the specimens, even if time and facilities for spreading the wings are not available, because pinned specimens are less liable to damage. Insert the pin vertically through the middle of the thorax (Figs. 71 and 120), leaving the body of the specimen about 13 mm (1/2 in.) below the top of the pin. Select a pin of the appropriate size. The scale of sizes for phalaenids and other average-sized stout moths is No. 3; for large geometrids and the smaller butterflies, No. 2; for small geometrids and large Microlepidoptera, Nos. 1 and 0; for most Microlepidoptera, No. 00; and for the smallest forms, No. 000. The last two sizes bend very easily; use them only when the larger sizes are not satisfactory. Never glue Lepidoptera to pins or points. For very small forms double mounts with minuten nadeln may be found necessary.

Fig. 120. A spread specimen (Lepidoptera).

For a permanent collection, Lepidoptera should be spread. Specimens, of course, must be flexible to be spread. The best results are obtained with freshly killed specimens. Green Geometridae must be spread when they are fresh.

The eggs of some Lepidoptera may be preserved dry. Many with delicate shells crumple when dried; keep this kind in alcohol. Most larvae should be killed in hot water to insure relaxed and undistorted specimens, and then preserved in alcohol. If you are going to study the internal anatomy, however, kill them by dipping them for a few seconds in boiling water or a hot fixative. Preservation in alcohol usually does not preserve colors, which should be recorded in notes and, if possible, by color photographs. Larval skins inflated with air or wax look very natural but their structure is hard to study. Save the molted skins and head capsules of larvae that are being reared and preserve them in alcohol. Also keep the pupal skins from which the adults have emerged. These and the cast-off larval skins should, of course, be clearly associated with the adults to which they belong, either by being placed on the same pin or by an appropriate system of labels.

A method of preparing the genitalia for study follows. Remove the abdomen from the specimen and treat it in caustic solution, as described previously. In the male, dissect the genitalia from the rest of the abdomen in 30% alcohol. Carefully pull the penis away from the rest of the genitalia. Make a fine slit at the anterior end of the aedeagus. Force alcohol through this from a hypodermic syringe to inflate the vesica. Transfer the aedeagus, with the attached vesica, to 95% alcohol, and keep the vesica inflated for a minute or two by pressure from the syringe, until it hardens in the inflated position. Transfer the rest of the genitalia to the 95% alcohol. Remove superfluous scales and hairs by careful brushing. Spread open the valvae and hold them in this position with a chip of glass until they harden. After about 20 min, transfer the genitalia to clove oil for clearing. If necessary they may be stained at this time by adding one part of saturated alcoholic safarinin to 100 parts of clove oil. Leave the structures in this solution for about half an hour, then transfer them to xylol, and mount them on a slide in a synthetic resinous mounting medium. Support the cover glass with chips of glass, to prevent damaging the genitalia as the mounting medium dries and shrinks. The procedure for preparing the female genitalia is almost the same as that for the male. Dissect the eighth and more posterior segments and the internal genital structures from the rest of the abdomen. Inflate the bursa copulatrix by means of a hypodermic syringe, inserting the needle through the ostium and down the ductus. Mount both male and female genitalia on slides with their ventral surface up. Try to keep the mounting position of your specimens uniform for easy comparison of different preparations.

Coleoptera

In Canada most species of beetles overwinter as adults and produce only one generation each year. Consequently, June is the best month for collecting; the adults of relatively few species are found in late July and early August.

Beetles occur in almost all habitats, and most species are secretive in their habits. For this reason, there is plenty of scope for ingenuity in devising new collecting methods.

Sweeping is a very productive collecting method. The vegetation in open woods or along the borders of woods and marshes often gives the most specimens. Sweeping during the last 2 hr of daylight on clear, warm, windless evenings is extremely productive; many species, especially those living on or in soil, are difficult to collect by any other method. Beating is also a good method of collecting many species.

Searching and handpicking is by far the most reliable method of determining food relationships of many species. Knowledge of such relationships is remarkably limited and erroneous. Many species, probably most of them, are restricted in their feeding. These remarks apply not only to leaf-eaters and borers, but to fungus- and sap-feeders, and predators.

Many aquatic species are taken by dredging pools and the margins of lakes and sluggish streams. Overhanging banks and water with emergent vegetation are often very productive. When the nature of the bottom permits, work the net rapidly, straining the water until the net becomes heavy with plant debris. Search the debris in the net, unless it can be spread out on bare sand, soil, or a white plastic groundsheet. Some collectors prefer the special dredging nets described on p. 61, but these nets are rarely more effective for collecting beetles than an ordinary aquatic net. Remove and examine stones and debris from swifter streams, or roll and rub the stones vigorously with your hand and allow the beetles to wash downstream into your net.

Many beetles may be found beneath stones, logs, loose bark, and other cover, and many can be sifted from leaf mold, moss, and other plant debris. The material from the hummocks of marshes is especially productive in early spring, when it contains large numbers of hibernating beetles. The drift of stream banks, lakeshores, and seashores is sometimes very productive, as is the debris washed out of irrigation ditches when they are first used in early spring. To collect beetles from such material, use a sifter (described on p. 54) to condense the material and then hand-sort it over a white plastic groundsheet or with a Berlese funnel. Decaying or fermenting vegetation, carrion, manure, and the soil beneath them often contain many beetles. They can be trapped with pitfalls. Many species are attracted to sap, and many others are attracted at night by lights, particularly those high in ultraviolet. The methods just described are some of the more productive collecting methods. But there are many other ways of collecting. For example, you can dig out the nests of ants and rodents, throw water on the banks of streams and ponds or shovel the banks into the water, tread down the emergent vegetation of marshes, search for nocturnal species with a flashlight, and devise other original methods.

The importance of using clean killing bottles has been stressed previously. Ethyl acetate is the most satisfactory killing agent for beetles, but you may prefer to use cyanide. Most beetles make better specimens if killed and stored in vials containing sawdust and ethyl acetate (p. 74). Many collectors prefer to preserve beetles dry, between layers of Cellucotton in pillboxes or wooden cigar boxes. Immature stages may be killed in hot water and preserved in alcohol. Most beetles other than those collected in alcohol or ethyl acetate require cleaning with ammonia before mounting.

Beetles may be pinned, pointed, or mounted on cards. Many collectors pin small beetles, which should be pointed. A general rule is: point all specimens less than 6 mm (1/4 in.) long, and unless they are very robust, point all specimens less than 9 mm (3/8 in.) long. If you follow this rule, you need pins of only three sizes: No. 1 for the smallest, No. 2 for medium, and No. 3 for the largest specimens and points.

Insert the pin near the base and inner edge of the right wing cover (as shown in Figs. 72 and 73). Many collectors tend to place the pin too far back or too far to the right. Locate the axis of the specimen at right angles to the pin, and very slightly elevate the anterior part of the specimen. Because protruding legs and antennae are likely to be broken, arrange these structures near the body. With freshly killed specimens, this can be done about 12 hr after they have been killed, when the muscles have lost their tone. Arrange relaxed specimens while you are pinning them.

Mount smaller specimens on points, with the tip of the point attached to the right side of the thorax, as described previously. For most beetles except lady beetles and tortoise beetles bend the tip of the point downward (Fig. 121).

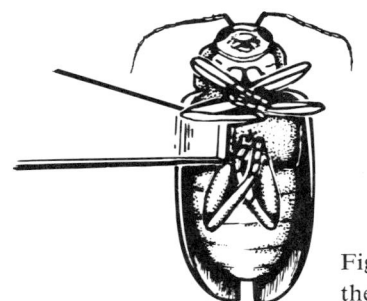

Fig. 121. A specimen (Coleoptera) showing the bending and position of the point.

Some beetles, particularly those with elongate and flexible bodies (such as Staphylinidae), should be mounted on cards (Fig. 122). Three sizes of cards are recommended. Cards should be of a good-quality paper (Bristol board); they are available from European supply houses. Place the beetle ventral side up on a white pinning board and hold it down by means of a device (Fig. 123) made from a cork stopper attached to a piece of strong celluloid, measuring about 3×16 mm (1/8 × 5/8 in.). Straighten the antennae and legs with a fine wet brush. Then turn the beetle over, place two or three small drops of glue in a line on a card of the appropriate size, and place the beetle on the glue with the same wet brush. Arrange the antennae and legs in the correct position. Always use the smallest size of card.

Fig. 122. A specimen (Coleoptera: Staphylinidae) mounted on a card showing the position of the pin and the specimen.

Fig. 123. A device for holding down a specimen mounted on a card.

Appendages should be arranged as shown in Fig. 122. This system of mounting elongate flexible specimens is much easier than pointing, and the body should remain level and the appendages straight. Subsequent study is easier and the specimens are more protected against damage. If you want to examine the undersides, it is easy to relax the specimens and remove them from the card by soaking them for a short time in warm or hot distilled water or an ammonia solution. Always use water-soluble glue. If you use water-insoluble glue, the specimen has to be soaked in a solvent, which usually causes hardening of the specimen. These specimens then become difficult to dissect or manipulate.

To restore pinned specimens that have become greasy, soak them overnight in benzene or carbon tetrachloride. Do not relax them. These liquids do not damage pins or labels. Relax pointed specimens that have become greasy, clean them in ammonia, and if you previously had used a water-soluble adhesive, remount them. Ammonia is particularly useful for quickly relaxing specimens that need remounting.

Strepsiptera

Stylops are rarely seen by most collectors. The best way to obtain them is to collect parasitized Hemiptera, Hymenoptera, or Orthoptera and to dissect out the female stylops or keep the hosts alive until the males emerge. Preserve stylops in alcohol.

Hymenoptera

This is one of the major orders of insects; it includes bees, wasps, ants, ichneumonflies, chalcids, gallflies, and their allies. These insects, like Hemiptera, are diversified in form, structure, and size. Consequently various methods of collecting and preserving must be used for the different groups.

Rearing is the best method of obtaining the parasitic forms (Ichneumonoidea and Microhymenoptera). The parasites are usually a by-product of rearing other insects, but preserve them carefully when you encounter them. Preserve the remains of the parasitized host and the parasite cocoon on the same pin as the parasite. It is important not to pull apart cocoon masses of gregarious species, because these masses often contain taxonomic characters. When adult parasites emerge, keep them alive for at least a day to harden and color.

Symphyta Sawflies, which are not strong fliers, are most active during sunny days and in hot, humid weather. Their main period of emergence is from early May to the end of June. They fly near marshes, in shrubby areas, in young forest stands and plantations, and in sheltered situations.

The eggs and larvae may be found by searching the food plants. The larvae appear about 2 weeks later than the adults. The larvae of most species are highly specific in their food requirements, though a few are more or less polyphagous. The cocoons may be found on the foliage, in galls, or on the ground close to the food plants.

It is desirable to rear sawflies, because the larvae of some species can be identified more easily than the adults, and when you have preserved material of both larvae and adults, the identification of others is much easier.

Preserve the cocoon from which an adult has emerged along with the adult, because it contains the last larval skin, which bears some significant taxonomic characters. Sawfly larvae are as easy to rear as are those of Lepidoptera. However, they require a high humidity for successful development. Most species spend the winter in the cocoon stage.

Adult sawflies may be netted as they fly, or they may be swept from foliage or flowers. Sweeping low ground plants in the morning is usually favorable for capturing newly emerged adults. Remove small specimens from the net with an aspirator. A Malaise trap may be used for collecting in favorable areas. Kill the adults with cyanide. Pin all specimens over 4 mm (1/6 in.) long directly through the thorax to the right of the midline (Fig. 78). Mount smaller specimens on points, taking care that the adhesive does not touch the wings. After the specimen has been pinned and while it is relaxed, push or blow the wings away from the body so that they do not cover the abdomen, spread out the legs and antennae away from the body, and exsert the saw of the female with a fine needle, carefully rearranging the saw sheaths in their natural positions, or exsert the male genitalia by gently squeezing the tip of the abdomen with fine forceps. With specimens that are to be pointed, this preparation of the genitalia must be done before they are affixed to the points.

Larvae may be preserved in alcohol. If they are dipped previously in a 1% solution of formaldehyde for 1 to 5 min they keep their colors better. The larval skins may be inflated. Use larvae of the last free-living instar rather than those of the final (prepupal) instar, which is passed in the cocoon, because the final instar lacks significant taxonomic characters of color and pattern. Make the opening at the anal end of the larva, through which the body contents are extruded and the skin is inflated, as small as possible, to preserve the specific characters of the supra-anal area. It is advisable to inflate some larvae from the head end.

Cocoons and samples of egg scars, galls, and injured leaves, twigs, or wood should be preserved dry. It is important that adults are associated by appropriate labels with preserved material of the early stages from which they have been reared.

Ichneumonoidea Ichneumon flies and braconids may be collected almost anywhere, but they are most abundant in moist environments in early summer. They may be taken individually while they are flying over foliage or grasses, or visiting flowers. Others are found running or flying near the ground or around tree trunks. In warm and dry regions, large numbers of Ichneumonoidea may be taken at light. In dormant seasons many Ichneumonoidea may be taken in hibernation.

Sweeping either from a definite host or at random yields good results including many smaller forms not otherwise taken. Do not make too many sweeps of the net before examining the contents, or the specimens become battered and dirty. Follow the method of sweeping described in the next section under Microhymenoptera. However, many Ichneumonoidea are too large to be taken by an aspirator, and these sweepings may be handled in a separator box.

Do not leave specimens too long in a cyanide killing bottle, because color changes are liable to occur. Pin and point them, depending on their size. Do not pin those under 1 cm (2/5 in.) long. Specimens about 1 cm long may be pinned only if the mesonotum is unusually wide (at least four times as wide as the diameter of the pin). The important thing to remember is to preserve the central part and at least one side of the mesonotum for taxonomic observation. Even specimens up to 2 cm (4/5 in.) long should be pointed if the thorax is too slim to be pinned. Affix the point to the specimen at the right side of the mesothorax (Fig. 124), and not on the ventral surface or the side propodeum.

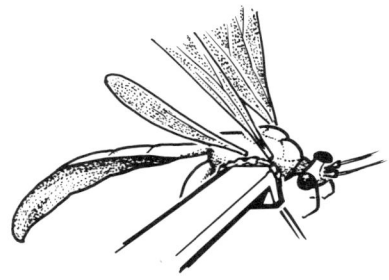

Fig. 124. A specimen (Hymenoptera: Ichneumonidae) showing the position of the point.

When you mount a specimen, tuck the legs under the body, but not so closely as to obscure the body or the legs. Fold the wings vertically over the back or spread them out sideways, but do not leave them folded flat over the propodeum and abdominal tergites, because these parts must be visible. Do not leave the antennae in an upright position beside the pin, or the tips eventually break off when the specimens are being handled; the same rule applies to the wings if they are long. Keep the abdomen straight, if possible, and do not bend it excessively upward or downward.

Because of the difficulty of handling and the risk of bending small pins, use only sizes Nos. 1, 2, and 3. Pin the specimens through the mesonotum, at the right of center. Insert the pin pointing a little backward so that the point pierces the mesosternum (Figs. 75 and 76). Never use a pinning block, because the pin always emerges through the space behind the forelegs, splitting the prothorax and head away from the body.

Do not discard a long series of gregarious or polyembryonic parasites. If the series is small (only 10 or 20 individuals), mount all of them. If, however, you want to save time or space, mount only part of a larger series (about 10 specimens, including both sexes if possible) and put the remaining material in a suitably sized gelatin capsule on a labeled pin along with the rest of the series. The specimens in the capsule may be cushioned with a minute wisp of cotton, if necessary.

Preserve Ichneumonoidea not mounted fresh by layering or in alcohol. Handle the specimens preserved in fluid by the following means. If the specimen has dried in a suitable mounting position and is clean, dehydrate it immediately by a quick soaking in 95% ethyl alcohol, and absorb the excess fluid on blotters after each bath. Then mount it in the usual way. If the specimen has dried in a position unsuitable for mounting, rearrange the parts in water before dehydrating it. Lift the specimen from the water and place it on a piece of blotter; the specimen should be lying on its side with its wings folded vertically, or upside down with its wings spread horizontally. After the paper and the specimen have dried for a few minutes, place them in 95% alcohol. It may be necessary to hold the wings in position with forceps for a few seconds till the alcohol penetrates and hardens them in position. Then proceed as previously described.

Microhymenoptera This group includes most of the chalcidoids, proctotrupoids, bethyloids, and parasitic cynipoids. These specimens are obtained by rearing, "berlesing," "malaising," suctioning, and pan trapping, but mainly by sweeping. During sweeping, usually enough debris accumulates to protect the insects from being battered. Both indiscriminate and selective sweeping of pure stands of vegetation are of value.

To separate the parasites from the finer debris in the bottom of the net, put the sweepings into jars with ethyl acetate or with cyanide. They can be examined later under a binocular stereomicroscope or strong hand lens. However, this method is slow. A second method is to catch the parasites with an aspirator as they crawl out of the net while you are gradually working through the debris searching for specimens. This method is inefficient because many specimens manage to escape. Nevertheless, by this method you can determine quickly the amount of material being collected and, by using a hand lens, the taxonomic groups represented. Furthermore, if you sweep selectively, there is no need to provide a bulky killing jar for each type of sweeping. A third method is to open the net with its sweepings near a window and then collect the winged forms from the glass (or you can use a car window); gather flightless species (which are rare) with an aspirator while you are inspecting the sweepings later. Regardless of what method you use to separate the specimens, examine the mesh of the net thoroughly afterwards for minute forms that have been trapped there.

Because the immature stages of insects and other arthropods may be parasitized, by collecting them you can accumulate clean parasite specimens with their host data. Unfortunately, the host is often unrecognizable from the stage obtained and adults of the host species fail to emerge. However, be sure to include all the available biological data on the label. When the host is being recorded make a note also of its stage or stages during the parasitism. For example, if you collect a caterpillar, and parasites emerge from its pupa, the parasite is a larval-pupal parasite and should be recorded as such. Preserve the remains of the host with the parasite because they may subsequently be used to identify the host, and to establish whether or not there was secondary parasitism.

Microhymenoptera must be mounted with the aid of a binocular microscope; otherwise many of the specimens need to be remounted for proper examination. A pair of very fine dissecting needles as well as a pair of fine brushes of good quality that have all but about 20 hairs removed (the hairs of cheap brushes curl readily) are useful for moving the appendages.

When possible, mount specimens that are going to be pointed soon after death. At that time, mounting is both easier and faster, because the appendages are easier to move and the wings do not need to be smoothed. If the insects have been layered, relax them immediately to prevent the appendages from breaking. If the specimens have been preserved in liquid, later on it will take you much longer to spread the wings, and also the color or sheen of the specimens may change. However, preservation of these insects in liquids, such as alcohol, is the easiest way of handling large numbers of specimens in the field. Moreover, during the preliminary handling and shipping, specimens in liquid are well protected from damage, except by the actual breakage of vials, or by complete evaporation; the latter hazard can be reduced by adding a little glycerin to the alcohol.

Sometimes you may collect a series of apparently only one sex of a species and you want to avoid mounting and labeling all that material, yet you are uncertain that only one species or sex is represented. In this case, point a few of the specimens and pack the rest securely with facial tissue in a gelatin capsule. In polyembryonic species, of course, do not mount many specimens, although for identification purposes the larval skin with its numerous parasitic pupal cells, or data labels stating that the specimens were polyembryonic, should accompany the specimens. If you use gelatin capsules instead of mounting, use only a few to supplement pointing, because any subsequent mounting of specimens or subsequent sorting is using the taxonomists's time rather than that of the collector, although the latter is primarily responsible for submitting material in reasonable condition for examination.

Chalcidoids and proctotrupoids are seldom large enough to be pinned. Most of them should be pointed. Slide mounts are rarely satisfactory. Only a few chalcidoids and proctotrupoids are too small to need the tip of the point bent down, a practice that encourages the use of too much adhesive. For the smaller species the point should end in a sharp fine tip rather than a coarse blunt tip; it is best to use a small amount of adhesive rather than a large drop that needs to be flattened out before pointing, because the glue spreads over too much of the body.

Many workers find the water-soluble adhesives preferable to other types of adhesives for pointing Microhymenoptera, because the one liquid dissolves the glue and relaxes the specimen, without the insect becoming brittle. Shellac gel and plastics can be used advantageously by deft workers accustomed to handling such small insects, thus allowing the appendages to be rearranged quickly and avoiding the hazard of repointing.

In a well-mounted specimen the antennae, wings, and legs are outstretched, not only to show these parts but also to display portions of the body that would otherwise be hidden.

Bring the antennae well forward and separate them from each other to show a clear silhouette of each segment, including the scape. This is essential for most specimens because the number, size, and shape of the various segments are used as the basis for identification. At times you may want to make a temporary mount of an antenna, especially in specimens of species that shrivel badly; if the antennae are outstretched, one of them can readily be detached without much chance of its being lost. After the examination, put the antenna on the upper side of the point by using a weak solution of a quick-drying water-soluble adhesive so that it can be adequately displayed without shriveling. In such cases, of course, leave one antenna attached to the head as a means of verification.

In relaxed specimens the wings usually extend upward or can readily be teased into this position. Point the specimen and, if necessary, immediately separate the wings from each other with a dissecting needle. Press the pair of wings nearest the point on to the adhesive so that they lie smoothly along the point; in this way you get a clear dorsal view of the taxonomically important propodeum, an area immediately in front of the abdomen (Fig. 125). If the specimen dies with its wings outstretched (as if it had been spread), you can point it in this position. If the wings are pointing downward, turn the insect upside down and put it on a slide, moisten the glass beneath the wings with water and press the wings evenly down. Then roll the body onto its left side and place the sticky point against the right side of the thorax.

To spread an insect, place the specimen on its left side on a slide with enough fluid to cover it. Hold the body still with either a brush or a needle,

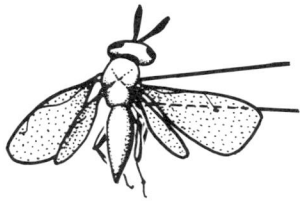

Fig. 125. A specimen (Hymenoptera: Chalcididae) showing the position of the point.

and use another brush or needle to straighten the antennae and legs. Then float and tease the four wings into an upright position, sometimes one on top of another but preferably with the two larger wings kept apart from each other and with their front edges toward the head. Draw off the fluid by pressing a piece of a common blotter lightly upon the wings, and remove any strands of blotting paper that may be left on the specimen. The antennae and legs may have become disarranged by the flow of liquid or by muscular contraction and they may need adjusting; if so, hold them in place by pressing them on a small remoistened part of the slide. Then place the point on the right side of the thorax and peel the insect from the slide toward the wings. Usually there is enough time for the point to be pinned and glued while the specimen is being dried; but, if you take too long, appendages other than the wing tips may need to be teased free from the slide. Examine the specimen for disarranged appendages and secure the right pair of wings evenly along the upper side of the point.

Apoidea, Sphecoidea, Vespoidea, Scolioidea Most of the remarks on Ichneumonoidea apply also to bees and wasps (Vespoidea, Scolioidea, and so on). Sweeping yields few of these insects. They are found in hot, barren, dry places, or at flowers and other sources of nectar, such as honeydew. Many species may be captured in a Malaise trap. These alert and vigorous insects can usually only be captured individually, and you must be very careful when handling them to avoid being stung.

Collect a large number of specimens, especially of small species, all of which look very much alike and may occur together. In social forms, nests may contain social parasites of very closely related species. In the genus *Polistes,* for example, many of the taxonomic problems will be solved only when complete colonies from many localities are available for study. Pin specimens through the thorax to the right of the midline (Fig. 77).

Formicidae Collect ants from their nests, which can be located by following the worker ants as they return to them. To obtain enough series (10 or more specimens) of each caste, examine the nest at intervals during the collecting season. Also preserve any other insects found in the nests. Pick up the ants with an aspirator and transfer them alive to screw-capped jars. A Berlese-type funnel is useful for separating them from debris in the jars. Preserve ants in alcohol. Pin some specimens, or, if they are under 6 mm (1/4 in.) long, point them.

Diptera

Such notorious pests as mosquitoes and black flies have made the two-winged flies a familiar, though unpopular, group of insects with the public. Therefore, few people consider collecting Diptera as a hobby. This is regrettable because most of the North American flies (about 16 000 known species) are harmless to humans, and many of them are attractive insects that perform beneficial roles in our environment.

In Canada, flies occur everywhere. They are the dominant insect group in the Arctic and they increase in numbers progressively southward. In southern Canada a few adult Diptera start flying as early as March, but the best times for collecting are from May to mid-July and from early September to mid-October or later, depending on the onset of cold weather. Late July through August is not usually productive except for a few sun-loving groups, such as Bombyliidae and Therevidae.

Habitats of Diptera cannot be discussed fully in this publication. For convenience, the collecting techniques used for adults and immature stages are treated separately. It is important to remember that the larvae of most flies require moisture, and the adults seldom move far from the larval habitat. Many flies are associated with specific plants or with phytophagous insects; therefore, to collect these flies, you can benefit greatly by having a good knowledge of your local flora.

Collecting adults   Use a standard net (*see* the section "Nets for Flying Insects," p. 14) to collect Diptera that are visiting flowers or resting on foliage. Also use a net for sampling hovering swarms of midges (Chironomidae) and other Nematocera. To catch strong-flying hoverers (Syrphidae, male Tabanidae, and so on), if possible, swing the net upward from below. For those that fly close to, and alight on, the ground (Miltogramminae, Therevidae, and so on), hold the net open and drop it over them so that they fly up into it. Once they are inside the net, secure the catch by twisting the handle and folding the bag over the rim. To escape from confinement, insects instinctively fly upward. If you examine the catch by looking down into the net from above, you will give many Diptera an opportunity to escape.

A sweep net (*see* the section "Sweeping," p. 33) of moderate weight is valuable for catching a variety of Diptera in large numbers. Keep the net dry and free from plant debris. Examine and remove the catch often (after every six or so sweeps). The body covering of Diptera (bristles, hairs, and pollen) is exceptionally delicate and its preservation is essential for accurate identification. Some groups of flies (Bombyliidae, Tipulidae, midges) rarely survive sweeping in good condition. See the section "Sweeping," p. 33, for various methods of emptying the net. If only a few insects are in the net, you can usually maneuver them directly into a killing tube after you have trapped them in the tip of your net. For collecting minute Diptera in habitats unsuitable for sweeping, such as the base of grass clumps, tree trunks, leaf litter, and lily pads, use an aspirator. Also use an aspirator for selective collecting of small midges and Acalyptratae on foliage, at a light, or trapped indoors on windows. Kill Diptera in cyanide or ethylene dichloride (*see* "Killing Agents and Killing Bottles," p. 73). Extract wingless forms and other small Diptera from soil, leaf litter, debris, and bird and mammal nests by "berlesing."

Fragile Nematocera, despite their small size, possess useful characters that are best seen in dry material. To prevent distortion through shriveling, kill, pin, and freeze-dry (p. 102, 156) a sample of each series (five or six of each sex). Put the rest of the series in alcohol while they are still alive, but if you need genitalia or other mounts, make these within a few months, because otherwise good slide material is hard to prepare.

Assembly sites Be alert to detect natural focal points of fly activity. Besides the nectaries of flowers, assembly sites include other plant and insect secretions, such as sap running from wounds in the bark of deciduous trees, especially elm and maple, and resin exuding from stumps of newly felled conifers. The opening leafbuds of poplars (*Populus* spp.) secrete a substance attractive to some Diptera, and the honeydew of aphids, especially on poplars, American green alder (*Alnus crispa*), and ironwood (*Ostrya virginiana*), attracts Tachinidae and other calyptrate flies.

Males of Chironomidae and other Nematocera hover in mating swarms over bushes or boulders, along margins of lakes, streams, and roads, using such sites as "swarm markers." In other families, such as Tabanidae, Syrphidae, and many Calyptratae, males congregate on bare hilltops and knolls, even in heavily forested country, where mating often takes place.

Predators such as Asilidae take up conspicuous positions on tips of bare twigs in clearings, evidently using them as lookouts when hunting their prey.

There are many other more familiar examples of Diptera assemblages, such as the concentration of blow flies around carrion.

Artificial light Porch lights of summer cabins beside lakes or streams attract many small aquatic Diptera, especially midges (Chironomidae). Collect them with an aspirator as they rest near the light. Small Acalyptratae also come to lights, sometimes in very large numbers during heavy rain. Trapping devices based on artificial light are described in the section "Lights and Light Traps," p. 15. The effectiveness of such traps depends largely on the selection of suitable sites for the operation.

Sunlight The tendency of Diptera to fly into warm buildings through doorways and then to seek to escape at windows provides an opportunity to collect interesting forms at windows of sheds, barns, greenhouses, sugar cabins, and similar places in open wooded country. Sometimes long series of normally scarce species are collected in this way.

Traps Natural baits consisting of food of larvae or adults can be used effectively for live trapping Diptera. For Tabanidae and Simuliidae, live animals or birds may be tethered in open-sided sheds or placed in special cages in suitable locations. For blow flies and dung flies such materials as raw meat, feces, or fermenting fruit and vegetables may be placed in wire-screen traps of various designs. However, sometimes such traps capture only female flies.

Humans attract some flies, but not all species of bloodsucking flies. Many Tabanidae prefer other mammalian hosts, and a large group of Simuliidae feeds exclusively on birds.

Chemicals Chemically produced odors simulating those of the natural food of insects seem to be, in general, unsuitable substitutes for the latter when used in traps. However, a lot of tests have been conducted on the effectiveness of collecting fruit flies in California. A list of chemicals and the species taken in each case has been published. Little work of this kind has been done in Canada.

Fresh malt extract, freshly diluted with water and sprayed on foliage with a nurseryman's sprayer, attracts Syrphidae, Tachinidae, and other flies. Apply this spray in early morning and repeat it hourly or more often. Make a fresh solution daily. It is not so attractive as aphid honeydew, and therefore is best used in late summer when honeydew is scarce. Also you can use fresh beer (instead of malt extract), diluted with an equal quantity of water, to attract the same groups.

Sugaring formulas commonly used for attracting Lepidoptera are usually not effective for Tachinidae, but they may attract muscids, sarcophagids, and calliphorids.

Unbaited traps Traps (Malaise trap and Herting tent trap), designed to catch insects in normal flight, are described on p. 23. In carefully chosen sites these traps are very effective for collecting Diptera. However, because flies are not easily restored after immersion in alcohol, replace the usual collecting container in a Malaise trap with a dry jar containing a killing agent. A small piece (6.5 cm^2, or 1 sq in.) cut from a dichlorvos strip, suspended in the head of the trap, and a similar block attached to the inside of the collecting jar have been used with good results. Fit a coarse-mesh screen across the entrance to the head of the trap as a barrier to heavy moths and beetles.

To prevent specimens from desiccating too quickly after death:

- Empty the trap several times daily in hot weather.
- Wrap the collecting container in foil.
- Place fresh leaves in the container to increase the humidity.
- Bury the container in the ground and connect it to the head of the trap by a vertical tube.

The last method keeps Diptera in excellent condition for up to 3 days, depending on soil temperatures. If, however, the buried container is cool enough that condensation occurs inside it, keep the flies from touching the walls by inserting a smaller copper-screen container inside the jar. If your catch is likely to be mainly Chironomidae, or other groups that are usually preserved in liquid, put some alcohol in your collecting jar. Use an aspirator to remove the specimens that require pinning from inside the trap.

Pan traps Pan traps (*see* also p. 30) are effective for collecting Diptera, especially in northern Canada. Specimens usually become greatly distended if left too long in a Formalin solution, but you can restore them by washing them in distilled water and drying them for at least 10 min on facial tissue. When pinned, they almost look like freshly caught material. If your specimens do not have to be mounted, wash them and transfer them to alcohol.

Preserving and mounting The general principles and methods of mounting and preserving specimens are given on p. 00. Most Diptera are hard-bodied and may be mounted on pins of the appropriate size. Besides pinned material, specimens preserved in alcohol are useful in many groups and are essential for the study of fragile Nematocera, such as Chironomidae, Ceratopogonidae, and Simuliidae, in which the bulk of the material is preserved in this way; only a few species of these families should be pinned (*see* a later section, "Freeze-drying," p. 156).

Pinning Pin specimens within a few hours of capture, while their internal parts are still soft and their appendages pliable. Insert the pin at the right of the center line of the thorax, near the base of the wing, and the point should emerge in front of the right mid coxa. Bunch and support drooping legs of Tipulidae during hardening by placing a piece of card about 9 mm (3/8 in.) below them on the pin. If specimens are too small and would be distorted by a size 0 pin, glue the right side of the thorax to the pin (size Nos. 0 or 1) with a small dab of shellac gel (Fig. 126). Whether the specimen is pierced or glued, insert the pin at right angles to the axis of the body and leave about one-third of its length protruding above the thorax (Fig. 127). Point wings backward, and slightly outward and upward. If the wings point forward or are drooping, move them to the desired position by slightly pressuring them against the base of the costa or on the thorax just above the root of the wing. Bunch the legs loosely below the body, without obscuring the pleural region, abdominal venter, or genitalia.

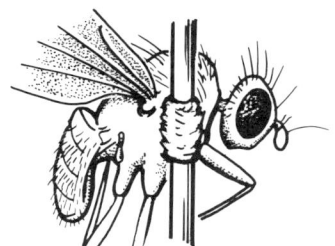

Fig. 126. A specimen (Diptera) glued to the side of a pin.

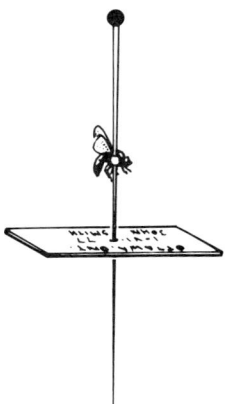

Fig. 127. A pinned specimen (Diptera) showing the position of the pin and the label.

Freeze-drying Small Nematocera and Brachycera (midges and empidids) are often too delicate to be mounted on pins in the usual way because of excessive shriveling, which alters their appearance and obscures their essential characters. However, if such specimens are mounted when they are fresh, then placed in a freezer and allowed to dry out slowly, shriveling is prevented and they continue to look lifelike. Although this method is especially useful for preserving delicate specimens, including reared material, it is also useful for all Diptera, because even robust forms undergo some shriveling of the abdomen and other parts. The length of time needed for this method of freeze-drying varies from about 3 weeks for small midges up to 6 months or longer for large flies.

Double mounts Double mounts made with card points and minuten pins are not recommended for small Diptera, because they take longer to prepare, are awkward to handle under a microscope, and do not increase the amount of viewable body surface. If you use card points, bend the tip of the point downward at a right angle and glue the right side of the thorax to the vertical surface of the point so that the specimens stay upright. Always insert minuten pins vertically at the right of the center line of the thorax. Use blocks made from polyethylene foam with minuten pins. The blocks should be as small as possible. The most satisfactory material for double mounting is a species of *Polyporus,* a bracket fungus, which is used extensively in Europe. It can be obtained in strips of suitable size from dealers in entomological equipment in England. It is of the right consistency to hold pins well and, because it is pure white, it enhances the appearance of the mounts. If specimens cannot be mounted when they are freshly caught, place them loosely in a small airtight container, with one or two fresh leaves, and store them overnight in a cool place, preferably a refrigerator. For longer storage, put them in a freezer in a tightly sealed container to prevent gradual drying.

Pinning from alcohol Store Diptera that were collected in alcohol (for example, from Malaise traps) in a freezer. Later, dry and pin this material by the following method. Transfer specimens to a 1:1 mixture of 70% ethyl alcohol and ethyl acetate. Mount larger specimens on pins while they are still in the liquid. Leave them for 24 hr, then transfer them to pure ethyl acetate to which two drops of ethylene glycol has been added. Leave them for 2 to 4 hr. Remove the specimens individually, place them on absorbent paper, and position the legs and wings correctly during drying. This can be done easily, because the specimens stay pliable for some time after their body surface has dried. When they are dry, glue small specimens to pins with shellac gel. Material that has been treated by this method usually dries without distortion, and with the pollen pattern restored and the wings clean and straight.

Dissected genitalia Store dissected genitalia in glycerin in microvials (p. 120) mounted beneath the specimen. Vials smaller than 3 mm (1/8 in.) internal diam are impractical, because the dissections are hard to remove and replace without smearing the sides of the vial.

Microscopic mounts Identification of small Nematocera (Chironomidae and Cecidomyiidae) is often dependent on the skill involved in the dissections and preparation of specimens for microscopic examination. Use fresh material if it is available. Relax pinned material in a moist chamber for 24 hr before you start. Material in alcohol is unsatisfactory if it has been stored longer than about 1 year. Use the following steps to prepare microscopic mounts.

- While the specimen is in alcohol, dissect the required parts under a stereomicroscope.
- Transfer dissections to glacial acetic acid for about 1 min.
- Leave the wings in acetic acid, transfer the other parts to a 5% solution of KOH in a test tube and heat it in a water bath for a few minutes until the internal tissues soften and dissolve. Do not overmacerate the parts.
- Transfer these parts back to the acetic acid, and gently expel all the dissolved tissue from the body cavities by pressure or with needles.
- Transfer all the parts, including the wings, to 2-propanol for about 1 min, then transfer them to fresh propanol in a dish with sloping sides.
- Gently pour cedarwood oil down the side of the dish so that it sinks below the propanol containing the dissections. Allow the dissections to sink gradually into the cedarwood oil at the bottom of the dish. Transfer the dissections to fresh cedarwood oil.
- Transfer each dissection, one at a time, to a small drop of Canada balsam on a microscope slide; use only enough balsam to prevent crushing the part with the cover slip. Add the cover slip and orientate the part by slightly moving the cover slip. If possible, mount all the parts on the same slide, using small circular cover slips.

Collecting and rearing immature stages Larvae of Diptera are found abundantly in still or running fresh water, living plant tissues, various soils, decomposing plants, carcasses, feces, and as internal parasites of other insects. They occur less commonly in brackish or saline water as predators of phytophagous insects, inquilines or commensals, and internal parasites of mammals and arthropods other than insects. Except in a few well-studied groups, such as biting flies, it is seldom possible to identify dipterous larvae beyond the family level. For this reason, you should always attempt to rear some insects in addition to collecting the immature stages. Many groups can be successfully reared without elaborate facilities, especially if you have collected nearly full-grown larvae. The principal factors to be considered in planning a rearing program are outlined on p. 70.

Aquatic larvae Separate the forms of aquatic larvae that require dissolved oxygen (for example, Chironomidae) from those breathing by direct contact with the air (Culicidae). To rear a single chironomid larva, put it in a corked vial about 2.5 cm (1 in.) in diam and containing water only 5 mm (1/5 in.) deep (to prevent asphyxiation). A mosquito larva in the same size of vial needs water at least 2.5 cm (1 in.) deep, and the vial should be covered with muslin or screening. Because of the importance of having a clean surface for both types of larvae, use distilled water and supply artificial food rather than relying on water from the original habitat. Some fish foods sold for freshwater aquariums are suitable diets for these larvae. Chironomidae found in running water may be reared in the same way as those in still water, but Simuliidae require a much closer simulation of their natural conditions. Galls and stemminers should be collected and examined in spring to ascertain the stage of development reached by the insect. Separate new galls from old ones that may persist on twigs for several seasons. Keep samples of live galls and make further collections from time to time until pupation takes place. Then isolate the galls until adults emerge. Take care to correctly associate immatures with adults.

Leafminers The larvae of leafminers (p. 46) develop rapidly and, in many cases, leave the mines to pupate in soil. By the time the mines are large enough to attract attention, they are often empty and, therefore, leaves should be collected soon after mining begins. Place each leaf in a stoppered vial that is just large enough to hold it without crushing it. Collect enough mines for dissection and recovery of exuviae of early instars. Press and mount examples of mined leaves as in a herbarium.

Soil-dwelling larvae Collect soil-dwelling larvae by the methods given on p. 00. Sifting, flotation, or both are perhaps most productive for soils, whereas Berlese funneling is best for debris. In rearing these larvae, it is important to approximate natural conditions and, unless a controlled-atmosphere room is available, use an insectary. Larvae from wet soils, such as Tabanidae, develop satisfactorily and are more easily monitored in wet moss or soft decomposing vegetation from the habitat than in the soil itself. For larvae from drier habitats, use sterile soil or a soil substitute only after the natural soil has been unsuccessful. Predatory larvae need a good source of food. *Tabanus* spp. eat fly maggots (*Musca*) and earthworms; do not leave uneaten fragments of worms, because they putrefy. Usually, tabanid pupae (and probably other soil-dwelling pupae) orientate themselves vertically, and if rearing conditions prevent this, adults may be deformed or fail to emerge. If necessary, properly orientate the pupae in the rearing container by hand.

Larvae in decaying vegetation For larvae in decaying vegetation (manure and compost heaps, leaf litter, and tree stumps), make a preliminary survey of the fauna by placing samples in polyethylene bags and identifying the adult insects that emerge. Likely predators (beetles, beetle larvae, and centipedes) can be recognized in this way, and then you can examine samples more carefully to eliminate these hazards and to search for pupae or puparia for rearing in isolation. Over several seasons the fauna can be gradually identified by elimination rather than by rearing individuals through all stages. The use of Berlese funnels speeds up analysis and often provides individual larvae for rearing.

Carrion and dung feeders These dung feeders can be reared with little difficulty in their natural media. However, with dung feeders as with those in decaying vegetation, predators must be eliminated. Blow flies oviposit readily in confinement and all stages including eggs are easy to obtain. Ground beef, liver, and pet foods are successful as diets for larvae of the blow fly.

Parasites First-stage larvae of Tachinidae may be dissected from gravid female flies. To obtain the later stages, rear the larvae of host insects, mainly Lepidoptera. Rear the host larvae, with their food plants, in polyethylene bags, in which you can maintain high humidity. Remove the puparia of emerging parasites, which are easy to see in the bags, and overwinter them in clean moist sand or sphagnum at about 2°C. Also overwinter hosts in diapause, with their unemerged parasites, in moist sand.

Aphid predators Aphid predators (Syrphidae and Chamaemyiidae) may be successfully reared if laboratory facilities are available to maintain stocks of host aphids under controlled conditions. Even without such facilities, useful data may be obtained by field observation and by collecting and rearing pupae. For preservation of larvae in best condition, kill them by immersing them in water heated almost to the boiling point. This hot water fully distends the body and also quickly deactivates enzymes that otherwise cause discoloration. Allow the water to cool slightly before transferring the larvae to alcohol (some forms stay more pliable in isopropyl alcohol). If larvae cannot be brought back to the laboratory alive, kill them by immersing them in KAAD (*see* "Formulas," p.169). Because of the bleaching effect of this solution, however, it is advisable to transfer the specimens to ethyl alcohol after 24 hr. Kill aquatic larvae, particularly Simuliidae, and preserve them by immersing them in 95% ethyl alcohol, because this insures distension of the mouthparts. Also, use 95% ethyl alcohol for preserving larvae killed in KAAD. Simuliid larvae for chromosome studies should be collected directly into freshly made Carnoy's fixative. In practice, you may find it is more convenient to collect all larvae into this fixative, transferring the surplus later to 95% alcohol. However, do not crowd too many larvae into the fixative (not more than about 10 larvae to 5 cm^3 (1 teaspoon) of liquid).

Larval and pupal exuviae Soften the larval and pupal exuviae that are obtained from rearings by washing them in a warm detergent solution and then preserving them in alcohol. The system of labeling such material must insure correct association of each adult with its exuviae. With isolated rearings from pupae collected in the field, you may prefer to mount the pupal case or puparium under the adult on the same pin (Fig. 128). Remove soil and other material adhering to the pupae by washing the pupae in detergent. Pin large pupae through the thorax or near the abdomen, and glue smaller specimens to a pin or a card point (Fig. 129) with shellac gel. Place skins smaller than about 1 cm (2/5 in.) long in microvials, either dry or in glycerin. In place of dry microvials, you may use small gelatin capsules, but these tend to become brittle with age and break when the pin is inserted.

Fig. 128. A method of mounting and associating the puparium with the adult.

Fig. 129. A method of mounting and associating the puparium with the adult.

Siphonaptera

Most species of fleas are ectoparasites of mammals; only a few species occur on birds.

The two principal methods of collecting fleas are by shooting or trapping the host animal and by searching the nests or bedding of the host. Because fleas are active and likely to leave the dead host animal, place freshly killed animals in paper or cloth bags as soon as possible after they die. When you use killing traps, especially on warm evenings, visit the trapline once or twice with a flashlight, and again at daybreak, or you will lose the fleas. Live traps obviate this disadvantage, but are large and bulky to handle.

Before you search for the fleas, quickly open the bags containing the dead animals and drop in a small wad of absorbent cotton soaked in chloroform. Close the bag again, and place it in a tin or jar for a few minutes to concentrate the vapors and kill the fleas. Then open the bag (tear open paper bags and flatten them, and inspect the folds at the bottom) and search the animals for fleas. Work vigorously through the fur with forceps to dislodge the fleas. Do this thoroughly or you will miss some specimens.

Place the fleas in a vial of alcohol and label the vial with the essential data. If there is any doubt about the species of host, preserve the animal whole, or, preferably prepare it as a standard museum "study skin."

The species differ in their seasons of adult activity so therefore collections should be made throughout the year.

Place nests of birds or mammals in a Berlese-type funnel, or tease them apart in a large white enamelled pan. Fleas are fairly conspicuous as they move about, and you can pick them up between hops with fine forceps moistened in alcohol. If you find flea larvae in the nest material, place them in a jar until they are reared. The importance of examining nests for fleas cannot be overemphasized, because these are often more productive of specimens than are the hosts themselves. This is particularly true of bird fleas. Some mammal fleas are almost entirely restricted to nests, and are therefore rare in collections, because this source of material is seldom exploited.

Fleas may be preserved indefinitely in alcohol, but they should ultimately be prepared as permanent microscope whole mounts. Clear them in cold 10% caustic potash, which usually takes about 24 hr, but sometimes longer with the darker and more heavily sclerotized species. It is seldom necessary to pierce specimens. Wash them for several hours in water, which may be slightly acidulated, and transfer them directly to 95% alcohol, and leave them overnight. Then, after an hour or two in absolute alcohol, place them for a few minutes in oil of wintergreen, and mount them in Canada balsam. However, the lactophenol process described for Mallophaga (see p.128) is a more satisfactory method of preparation. By this method, the fleas are not stained but are placed, one to a slide, with the head to the right, and the legs pointing away from the slide (so that the insect will appear right side up and facing left when viewed under the compound microscope). Affix two labels to each slide, the one on the left with all the collection data, and the other with the name of the species after it has been identified.

Acari (mites)

The mites, as a group, rival the insects in terms of diversity of form, morphological structure and habits, and of occurring everywhere. Ticks (Ixodides or Metastigmata), as one suborder of Acari, are discussed separately in the next section. Mites are second only to the insects in number of families, genera, and species, and a few acarologists now believe that they will be found to equal or surpass the insects in this respect. They are found throughout the world in terrestrial and aquatic environments. The aquatic forms occur in fresh and salt water, including the bottoms of ocean shallows and ocean depths, which have not been colonized successfully by insects. They abound in soil and organic debris, where they normally outnumber other arthropods. Many of the free-living mites are predators on other small arthropods. Many others are fungivorous or feed on vascular plants. A great variety of mites are parasitic during part or all of their life cycle; most of these are ectoparasites, having both vertebrates and invertebrates (including many insects) as their hosts. Many other mites, particularly in the order Cryptostigmata or Oribatei, are scavengers. Perhaps 20 000 species of mites, representing nearly 300 families, are known from North America; this represents at best only 10% of the species of mites considered to exist in this region.

The thorough collecting of mites in a given area depends on the recognition and selection of various collection sites representing the different habitats of the area in which mites are likely to occur. Ectoparasitic acari, for example, may be found in nests or along runways of their hosts, but they would rarely be found otherwise apart from their hosts. Free-living terrestrial forms mostly prefer substrates protected from sunlight and wind, with sufficient moisture for maintaining a satisfactory water balance. However, although the mites that occur in exposed or arid locations may be relatively sparse in numbers and kinds, they may be xeric forms that are found only in such places.

The methods of collecting mites are nearly as varied as the habitats in which they occur. No single method is satisfactory for collecting various species of mites. Many specialized methods have been devised, because most mites are too small to be handled by entomological techniques, yet too large to be managed by microbiological procedures.

The simple but somewhat time-consuming method of collecting mites directly can be very rewarding. When you see them with the naked eye or with the aid of a hand lens, use either a fine artist's brush moistened with alcohol or a bulb aspirator to pick them up. The bulb type of aspirator is described in the section "Collecting Aquatic Insects and Mites," p. 57. By this means, the inhabitants of rock fissures, walls, bark, and similar habitats can be captured. This method gives you the greatest opportunity to record observational data on the ecological niche and sometimes even on the feeding habits of the mites. It is helpful also when you are searching for parasitic mites, many of which are characteristically localized on their hosts and do not come off the fur of mammals or the feathers of birds by the usual methods of combing or shaking the host specimen. Many forms of plant mites, both plant-feeders and predators, can be collected by direct methods, particularly if you take the infested parts of the plant to the laboratory so that the mites can be removed with the aid of a binocular stereomicroscope.

Direct methods are impractical and inadequate for collecting mites from many habitats. Litter-dwelling and nidicolous (sharing a nest with another kind of animal) forms hide themselves among the multitude of organic fragments of their niches. Further, a collector cannot hope to recognize, distinguish, and pick out representatives of each of the many different species that commonly coexist in such habitats. The solution is to collect samples of the different habitats, rather than the mites, for subsequent separation of the occupants hiding within. The degree of precision, or purity, with which a substrate is selected also gives you a lot of information about the ecological nature of the habitat in which the mites live. This is a great advantage over the entomological techniques of Malaise traps, light traps, and fly-by traps, which gather insects from various unknown habitats in a general vicinity. Extraction methods are used for removing and collecting arthropods from habitat samples. They are suitable for qualitative faunistic analyses of habitats, and they are also adaptable to standardized sampling and replicating, if necessary, for statistical and synecological studies. Extraction methods are of two types: dynamic and mechanical. Dynamic methods, with the use of equipment such as the dry funnel of the Berlese apparatus, depend on the activity of the arthropods and their movement out of a habitat sample containing them in response to a stimulus (usually repellent in nature). With mechanical methods, such as flotation, suction, and brushing, the arthropod's role is passive, and separation depends on such factors as the composition and density of the habitat sample, the size and density of the arthropods, and the properties of the arthropod's integument.

Phytophagous, fungivorous, predaceous, and saprophytic mites from various habitats (litter, soil, moss, fungi, higher plants, nests, on and under bark, and in decaying wood or other organic material) can be extracted in large numbers by means of a modified Berlese funnel. The advantages of the Berlese funnel over other extracting devices are that its extracting action is automatic, called an autosegregator; the extracted arthropods are recovered intact, and may be collected alive for rearing or preserving in liquid; and it is the best means of extracting arthropods from litter and other substrates high in organic matter. One disadvantage of the funnel method is that it depends on the mobility of the animals and that therefore the inert forms (eggs, hypopi, pharate, and teneral adults) cannot be collected. Also, it depends on the responses of a wide variety of arthropods that have behavioral and physiological differences to a single repellent stimulus so that there is a bias against forms that do not react favorably to the particular stimulus. In the latter respect, to many arthropods, desiccation from a source of heat appears to be a better stimulant than any of the chemical repellents. The description and operation of the Berlese funnel is given on p. 47.

Mites in soil, humus, and matted ground vegetation can also be extracted by one of the various flotation methods. The advantages of these methods are that inactive as well as active forms of arthropods are extracted, and a more complete and unbiased extraction is attainable because the arthropods do not need to have a reaction to a stimulus. The effectiveness of these methods does not depend on the condition of the sample substrate (whether it is moist or disturbed) and is superior to that of funnels for extracting arthropods from soils with a high clay content. However, these methods are tedious and definitely not automatic, and the mechanical processes of washing and sieving sometimes damage fragile specimens. Also, these methods do not lend themselves to extracting arthropods alive for purposes of rearing. Finally, mites in substrates containing much organic material do not separate well, because not all of the debris is disposed of. It is advisable, therefore, to substitute a flotation method in place of a Berlese apparatus for extracting mites only if improved collecting results are anticipated. For example, collectors have failed to collect rhodacarid mites from heavy loam soils with the use of desiccating funnels, but by using a flotation extraction they have found that these mites are the dominant group of Mesostigmata in the loam community. The flotation method is described in the section on collecting methods (*see* "Collecting and Extracting Burrowing and Boring Arthropods," p. 56).

Place habitat samples collected for extraction in cloth or heavy paper bags, which allow for air exchange in the contents. In this way, samples can usually be maintained for several days before extracting, without adverse results. The effects of keeping samples in different kinds of bags is given in the section on collecting methods (*see* "Collecting and Extracting Arthropods from Debris," p. 47). Affix an appropriate collection data label for the sample to or in each bag.

To collect free-living mites on plants, whether they are phytophagous, fungivorous, predaceous, or saprophagous, beat the foliage of branches against the screened surface of a white tray. The mites are easy to see as tiny moving specks on the tray surface, where they may be collected with an aspirator or a fine brush. Beating funnels, powered devices for brushing mites from leaves, and motorized suction devices can also be used for collecting mites from plants, although the collector knows less about the results of his catch until it has been examined later in the laboratory. The description and operation of the equipment used in these methods, and references to motorized devices are given in the section on collecting methods (*see* "Beating," p. 34).

Several species of mites occur with insects that tunnel under bark or into wood or into the woody conks of bracket fungi. These may be collected directly with a brush or aspirator, because the tunnels can be exposed by prying the bark with a knife, small crowbar, or axe; or they may be extracted by placing small broken-up pieces of the tunneled habitat in a Berlese funnel.

Aquatic mites, also, can be collected in two ways. Free-swimming forms in ponds and slowly moving streams can be collected directly with a small dip net. The more conspicuous ectoparasitic larval mites on many kinds of adults of aquatic insects can also be collected with a small net when you see them with your naked eye or with the aid of a hand lens. However, you need to use several devices and procedures to collect many other kinds of water mites that are hidden in other habitats such as in weedy areas along lake shorelines, or in algal beds, coral formations, and deep water or on the bottoms of deep lakes and ocean floors. The use of such equipment is described in the section on collecting methods (*see* "Collecting Aquatic Insects and Mites," p. 57).

There are many ways of locating and collecting the various kinds of mites that are facultative and obligate parasites of birds, mammals, reptiles, insects, arachnids, myriapods, and other animals. Direct examination of the animal host, particularly with the aid of a stereomicroscope, is rewarding, especially because further data can be gathered, such as the specific or preferred locations of the mite on the host, the method of attachment, and the degree of injury to the host. Procedures for examining dead and living vertebrate hosts for mites and special methods for collecting unattached parasitic mites are given in the section on collecting methods (*see* "Collecting Ectoparasites of Vertebrates," p. 67). Descriptions of procedures are given in "Collecting Mites Associated with Invertebrates," p. 69.

The rearing and culturing of mites is a subject that is too extensive to be dealt with in this publication. The methods depend in large part on the feeding habits and natural habitat of the mites. The general principles of successfully rearing arthropods are given in a separate section of this book (*see* "Rearing," p. 70).

To kill and preserve most mites, put them in 70–80% ethyl alcohol and add 2–5% glycerin, which prevents the specimens from drying out if the alcohol evaporates. Restore dried-out specimens to a pliable, unshriveled condition by soaking them in warm 50–80% lactic acid (which also acts as a clearing agent) for a couple of days, and then placing them in alcohol again. Some workers have kept cleared specimens indefinitely in lactic acid, which acts as a preservative and as a medium in which the specimens may be observed on temporary slide mounts. Oudemans' fluid (*see* "Formulas" at the end of this publication) is preferred by some workers as a preservative, because it kills the specimens with their appendages extended, and then they are easier to mount on slides later. Preserve eriophyid mites dry, in situ on their plant hosts; and as with other dried mites you can mount them directly into Hoyer's medium to study them. When you collect ectoparasitic mites that have become darkened with engorged blood or are heavily sclerotized, first, put them in Koenicke's fluid (*see* "Formulas") or in a 2% aqueous solution of chloral hydrate to start clearing them, and later transfer them to alcohol for indefinite storage.

Most mites should be cleared and mounted on microslides before you can study them under a compound microscope. Make whole mounts for most mites, and make dissected preparations of some of the larger, heavily sclerotized forms of Mesostigmata and Cryptostigmata (Oribatei) and of some of the large soft-bodied forms of Parasitengona and Hydrachnellae (water mites).

Many mites do not need to be cleared before mounting, because of the clearing action of lactic acid as a temporary mounting medium or the chloral hydrate ingredient of the semipermanent Berlese-type mounting media. You can control the degree of clearing in these instances by the length of time you leave the slide preparation in the warming oven. Some additional clearing usually takes place in the mounting medium after the period of heating, but usually not enough to overly clear the specimens. Heavily sclerotized mites often need to be cleared by means of a separate procedure: soak them in 50–100% lactic acid (the concentration depending on the degree of sclerotization) or in a slightly more potent lactophenol for 2 to 5 days, or for a shorter time if the solution is kept warm. Puncture larger specimens along the lateral or terminal body wall so that the lactophenol penetrates more easily into the body cavity. Also, similarly puncture mites engorged with blood or filled with dark pigment and then gently squeeze them to remove some of these substances from their bodies; squeeze them gently again after the specimens have soaked for a day. Dried mites also can be restored to a pliable condition for mounting by first immersing them in lactophenol or in 50–80% lactic acid for 1 or 2 days. Nesbitt's fluid is a more powerful chloral hydrate clearing agent, which is recommended for heavily sclerotized specimens or specimens stored for prolonged periods in alcohol that do not clear sufficiently in other solutions. Do not leave specimens in warmed (50°C) Nesbitt's fluid for longer than 1 hr without checking them, and, in any event, for no longer than a day, to avoid weakening and damaging the cuticle.

There are two methods that have been found effective for studying mites on slides. Make temporary preparations in lactic acid and then use a cavity slide or make semipermanent preparations in a water-soluble gum-chloral mounting medium such as Hoyer's. These methods are described in the section "Microscopical Preparations," p. 102. Permanent resinous mounts are not recommended. To make semipermanent preparations when more than one specimen of a species is available for mounting, place one specimen with the dorsal side up, and the other with the ventral side up. Lateral mounts of mites or their parts are also needed for studying some taxonomically important structures, such as the chelicerae of many Mesostigmata, the male aedeagus of many Prostigmata and Astigmata, and the lateral plating, setation, and other cuticular structures of most Cryptostigmata (Oribatei). Mount only one mite on each slide. Use cover slips 12–15 mm (1/2–3/5 in.) in diam and No. 0 or No. 1 thick, to allow for study with oil immersion objectives. After you place the cover slip on the preparation, immediately apply moderate heat, just enough heat that the medium does not start to bubble. The heat will spread the medium so that it flows quickly under all reaches of the cover slip and helps extend the appendages and reduce the cuticular wrinkling and other irregularities produced by the semipermanent mounting technique. The subsequent procedures of gently warming semipermanent preparations to set and further clear them, and of sealing them with a water-insoluble compound around the periphery of the cover slip to increase their permanency are also described in the section on preserving and mounting (*see* "Microscopical Preparations," p. 102). The procedure for retrieving and remounting mites from deteriorated semipermanent preparations is also described in the same section.

The storage and care of mites in a liquid collection and as slide preparations and the study of mites from these collections are described in the sections under the storage and care of collections (*see* "Storage of a Collection Preserved in Liquid," p. 116; "Storage of Microscope Slides," p. 116).

Ixodides

Ticks are mostly ectoparasitic on mammals, although a few species occur on birds. Except the "soft ticks," which may feed and molt more often, species found in Canada have three feeding stages: the larva or "seed," the nymph, and the adult, each of which must obtain a blood meal to complete the life cycle. Usually, ticks drop off between meals to molt, and must again seek a host, not necessarily of the same species as the preceding one.

Ticks occur in nature in three principal situations, each requiring a different collecting technique: (*a*) on host animals, (*b*) in the host's nest or bedding, and (*c*) free-living.

(*a*) When the host animal is shot or trapped, place it in a bag or other container to prevent losing the ticks and other parasites. Ticks usually release their mouthparts from an animal that has been dead for a time, and they may be found wandering in the container or on the fur or feathers of the host. Search the animal carefully, especially around the neck and ears, with forceps for removing attached ticks. Ticks are tough and can usually be removed without damaging or losing mouthparts; grasp them well forward, and pull with a gentle twisting motion. Set aside engorged nymphs in rearing tubes if you want to search for the adults, or you can place the specimens in preservative, usually alcohol, with a little glycerin added. Killing in hot alcohol has the advantage of making the ticks extend their legs.

(*b*) Examine the nests of mammals and birds for ticks. It is sometimes possible by this means to find the males of species, which otherwise are difficult to obtain. A Berlese funnel may be used, but ticks are usually found readily if the nest material is teased apart in a white enameled pan under a strong light.

(*c*) The adults of a few species may be collected in numbers in a free state in locations where they are perched, waiting for a passing host. The locations vary with the species, as does the season of activity, although most species may be collected in the early spring. The ticks are usually perched on stems or blades of grass, or on low-lying bushes (*see* p. 68).

Keep ticks in alcohol, because the specimens are usually easy to identify when they have been preserved in alcohol. To prepare slides, use the standard potassium hydroxide – balsam technique. Make a small slit in one side of the abdomen to allow the fluids to penetrate and macerate for easy removal of the body contents. With thick specimens you may have to prepare a cell on the slide.

Araneae

Collect spiders by searching, sweeping, beating, sifting, pitfall trapping, or vacuuming. A useful aspirator (Fig. 31) that permits examination of the living specimens and subsequent transfer to individual vials or rearing cages consists of a 30 cm (12 in.) length of glass tubing (about 1 cm, or 2/5 in., in diam) to which is fitted a rubber suction tube about 70 cm (28 in.) long. The joint holds a piece of fine screen. Catching is improved by bending the tip of the glass tube at an angle of about 30°. A vial-type aspirator is useful if the entire collection is going to be put directly into preservative. Use a small amount of alcohol to calm the spiders, because they often attack and injure each other, and clog the vial with silk when they are crowded together alive.

Sweeping and beating are mainly used for collecting spiders from trees, shrubs, or herbs, and particularly for web-builders such as Theridiidae, Araneidae, Tetragnathidae, and Dictynidae and hunters of the families Clubionidae, Salticidae, and certain genera of crab spiders. Sifting yields litter inhabitants. Pitfall traps are best for the larger Lycosidae and Thomisidae, but they also yield large numbers of certain Clubionidae, Gnaphosidae, Linyphiidae, Erigonidae, and Agelenidae. More males than females of some groups are caught in pitfall traps. A common pitfall trap consists of a container such as a white enamel saucepan 30 cm (12 in.) in diam set into the ground with its lip level with the surface. If you use a dry pan (for live specimens), place a screen in the bottom of the pan for drainage and a handful of leaves or litter for hiding places. If you use a wet pan, pour into it some alcohol, and add water (to which a little detergent has been added to ensure wetting) or ethylene glycol. A raised lid, which may be camouflaged, shields the contents from rain. Empty the glycol traps at intervals of not more than 2 or 3 weeks, because the fluid is hygroscopic and the specimens are liable to decay. Traps containing other fluids should be tended daily. Use from two to six traps, set a few metres apart, in each location.

A vacuum collector mechanizes the collection of spiders from litter and herbs. It is operated by a gasoline motor (which can be carried on your back or pushed on wheels) and it has a flexible hose. The large ground-running spiders usually escape capture unless you set up an enclosure of some kind. To separate the specimens from the debris, use an aspirator on a sheet or a Berlese funnel.

Formulas

AGA solution — parts
    commercial ethyl alcohol — 8
    distilled water — 5
    glycerin — 1
    glacial acetic acid — 1

Acetoalcohol mixture — parts
    commercial ethyl alcohol — 5
    glacial acetic acid saturated with corrosive sublimate — 5

Alcohol (ethyl or methyl)

Take the amount of 95% alcohol (that is, commercial ethyl alcohol) equal to the concentration desired, and add water until the total volume is 95 cm^3. For example, to make 75% alcohol, take 75 cm^3 of commercial ethyl alcohol and add 20 cm^3 of water, making 95 cm^3. Add 2 cm^3 of glycerin if desired.

Andre's fluid
    glacial acetic acid — 50 cm^3
    chloral hydrate — 50 g
    distilled water — 50 cm^3

Barber's fluid — parts
    commercial ethyl alcohol — 53
    water — 49
    ethyl acetate (acetic ether) — 19
    benzol (benzene) — 7

Barr solution
    saturated solution of Eosine Y in glycerin–alcohol
    Eosine Y in commercial ethyl alcohol — 10 cm^3
    glycerin — 10 cm^3

Berlese's medium
    chloral hydrate — 160 g
    gum arabic — 15 g
    glucose syrup — 10 g
    acetic acid — 5 g
    distilled water — 20 cm^3

Brasil's fluid
    ethyl alcohol (80%) — 150 cm^3
    picric acid — 1 g
    Formalin (40% formaldehyde) — 70 cm^3
    glacial acetic acid — 15 cm^3

Carbolfuchsin stain
    saturated solution of basic Fuchsine — 10 cm^3
    5% solution of carbolic acid in distilled water — 100 cm^3

Carnoy's fixative | parts
    commercial ethyl alcohol | 3
    glacial acetic acid | 1

Carnoy's fluid (modified)
    isopropyl alcohol .... 60 cm^3
    chloroform .... 30 cm^3
    formic acid (90%) .... 10 cm^3

Caustic potash (KOH) 10% solution
    Dissolve 5 g of pellets or a stick of potassium hydroxide in 50 cm^3 of distilled water.

Chloral hydrate clearing solution
    chloral hydrate .... 40 g
    water .... 25 cm^3
    hydrochloric acid .... 2.5 cm^3

de Faure's medium
    gum arabic .... 30 g
    chloral hydrate .... 50 g
    chlorhydrate of cocaine .... 0.5 g
    glycerin .... 20 cm^3
    distilled water .... 50 cm^3
    Mix and filter.

Formal-acetic mixture | parts
    Formalin (40% formaldehyde) 10
    glacial acetic acid 10
    water 80

Hoyer's medium
    gum arabic (crystals) .... 30 g
    distilled water .... 50 g
    chloral hydrate .... 200 g
    glycerin .... 20 g

    Dissolve the gum arabic in the distilled water at room temperature. Add the chloral hydrate and leave the mixture for a day or two until all solids have dissolved. Add the glycerin. Filter through glass wool. Store in a bottle with a glass stopper.

Kerosine – acetic acid – dioxane (KAAD) solution | parts
    kerosine 1
    commercial ethyl alcohol 10
    glacial acetic acid 2
    dioxane 1

    For very soft-bodied larvae, use half as much kerosine, or less. Dioxane may be omitted.

Kahle's fluid | parts
    commercial ethyl alcohol 15
    distilled water 30

Formalin (40% formaldehyde)	6
glacial acetic acid	1

Koenicke's fluid
glacial acetic acid	10 cm^3
glycerin	50 cm^3
water	40 cm^3

Lactophenol – Methyl Cotton Blue stain
phenol	500 cm^3
glycerin	1000 cm^3
distilled water	500 cm^3
Methyl Cotton Blue	0.0025–0.01%

Dissolve Methyl Cotton Blue in the water before adding it to lactophenol.

Nesbitt's fluid
chloral hydrate	40 g
distilled water	25 cm^3
hydrochloric acid (1 N)	2.5 cm^3

Oudemans' fluid
70% ethyl alcohol	87 cm^3
glacial acetic acid	8 cm^3
glycerin	5 cm^3

Pan-trap solution
ethylene glycol	200 cm^3
Formalin (40% formaldehyde)	2 cm^3
liquid detergent	8–10 drops
water	800 cm^3

Pempel's fluid
	parts
glacial acetic acid	4
Formalin (40% formaldehyde)	6
distilled water	30
commercial ethyl alcohol	15

Polyvinyl lactophenol
commercial polyvinyl alcohol (Elvonol, type A, 51.A.05)	2.5 g
lactophenol solution (45 g phenol detached crystals in 45 cm^3 of lactic acid)	30 cm^3

Solutions of lower refractive indexes may be obtained by dissolving the polyvinyl alcohol in 10–15 cm^3 of distilled water.

Shellac gel

Boil 150 cm^3 of white shellac for 20 min. Add 10 cm^3 of 70% ethyl alcohol and boil it for 5 min. Cool the mixture rapidly by pouring it into vials in cold water. The resulting medium should have the consistency of petroleum jelly.

References

Anderson, R. M. 1948. Methods of collecting and preserving vertebrate animals. Natl. Mus. Can. Bull. 69(18):162.

Aucamp, J. L., and Ryke, P. A. J. 1965. The efficiency of an improved grease film extractor in stratification studies of soil microarthropods. S. Afr. J. Sci. 61:276-279.

Baker, J. R. 1958. Principles of biological microtechnique. Methuen, London. 357 pp.

Barr, D. 1973. Methods for the collection, preservation, and study of water mites (Acari: Parasitengona). R. Ont. Mus. Life Sci. Misc. Publ. 28 pp.

Beaudry, J. R. 1954. A simplification of Hubbell's method for trapping and preserving specimens of *Ceuthophilus* (Orthoptera, Gryllacrididae). Can. Entomol. 86:121-122.

Borror, D. J., DeLong, D. M., and Triplehorn, C. A. 1976. An introduction to the study of insects. 4th ed. Holt, Rinehart, and Winston Publishers, New York. 852 pp.

Borutzky, E. B. 1955. A new trap for the quantitative estimation of emerging chironomids. Tr. Vses. Gidrobiol. O-Va. 6:223-226.

Burton, W., and Flannagan, J. F. 1973. An improved Ekman-type grab. J. Fish. Res. Board Can. 30(2):287-290.

Chant, D. A. 1962. A brushing method for collecting mites and small insects from leaves. Pages 222-225 *in* Progress in soil zoology. Vol. 1. Butterworth & Co. Ltd., London.

Corbet, P. S. 1965. An insect emergence trap for quantitative studies in shallow ponds. Can. Entomol. 97:845-848.

De Giusti, D. L., and Ezman, L. 1955. Two methods for serial sectioning of arthropods and insects. Trans. Am. Microsc. Soc. 74:197-201.

Dominick, R. B. 1972. Practical freeze-drying and vacuum dehydration of caterpillars. J. Lepid. Soc. 26(2):69-79.

Edmondson, W. T., and Winberg, G. G. 1971. A manual on methods for the assessment of secondary productivity in fresh waters. IBP Handbook No. 17, Int. Biol. Prog. London. Blackwell Scientific Publications, London. 358 pp.

Evans, G. O., Wheals, J. G., and Macfarlane, D. 1961. The terrestrial Acari of the British Isles. Vol. 1. Br. Mus. Nat. Hist., London. 219 pp.

Foote, R. H., and Blanc, F. L. 1963. The fruit flies or Tephritidae of California. Bull. Calif. Insect Surv. 6:117 pp.

Grandjean, F. 1949. Observation et conservation des très petits arthropodes. Bull. Mus. Natl. Hist. Nat. Ser. 2, 21:363-370.

Gressitt, J. L., and Gressitt, M. K. 1962. An improved Malaise trap. Pac. Insects 4:87-90.

Haarløv, N., and Weis-Fogh, T. 1955. A microscopical technique for studying the undisturbed texture of soils. Pages 429-432 *in* D. K. McE. Kevan, ed. Soil zoology. Butterworth & Co. Ltd., London.

Hardwick, D. F. 1950. Preparation of slide mounts of Lepidopterous genitalia. Can. Entomol. 82:231-235.

Hardwick, D. F. 1968. A brief review of the principles of light trap design with a description of an efficient trap for collecting noctuid moths. J. Lepid. Soc. 22(2):65-75.

Haufe, W. O., and Burgess, L. 1960. Design and efficiency of mosquito traps based on visual response to patterns. Can. Entomol. 92:124-140.

Herting, B. 1969. Tent window traps for collecting tachinids (Dipt.) at Delémont, Switzerland. Commonw. Inst. Biol. Control Tech. Bull. No. 12:1-19.

Hopkins, G. H. E. 1949. The host associations of the lice of mammals. Proc. Zool. Soc. (London), Part II:387-604.

Hubbell, T. H. 1936. A monographic revision of the genus *Ceuthophilus* (Orthoptera, Gryllacrididae, Rhaphidophorinae) Univ. Fla. Biol. Sci. Ser. II(1):12-14.

Hurd, P. D. 1954. "Myiasis" resulting from the use of the aspirator method in the collection of insects. Science (Wash., D.C.) 119:814-815.

Kempson, D., et al. 1962. A new extractor for woodland litter. Pedobiologia 3:1-21.

Lewis, G. G. 1965. A new technique for spreading minute moths. J. Lepid. Soc. 19(2):115-116.

Lipovsky, L. J. 1951. A washing method of ectoparasites recovery with particular reference to chiggers (Acarina – Trombiculidae). J. Kans. Entomol. Soc. 24:151-156.

MacFadyen, A. 1962. Soil arthropod sampling. Adv. Ecol. Res. 1:1-34.

Malaise, R. 1937. A new insect trap. Entomol. Tidskr. 58:148-160.

Mason, W. R. M. 1974. Shipping alcohol collections in plastic bags. Proc. Entomol. Soc. Wash. 76(2):229-230.

Mitchell, R. D., and Cook, D. R. 1952. The preservation and mounting of water-mites. Turtox News 30(9):1-4.

Mundie, J. H. 1956. Emergence traps for aquatic insects. Mitt. Int. Ver. Theor. Angew. Limnol. No. 7:1-13.

Murphy, P. W. 1962. Extraction methods for soil animals. II. Mechanical methods. Pages 115-155 *in* Progress in soil zoology. Vol. 11. Butterworth & Co. Ltd., London.

Newell, I. 1955. An autosegregator for use in collecting soil-inhabiting arthropods. Trans. Am. Microsc. Soc. 74:389-392.

Newell, I. M. 1967. Abyssal Halacaridae (Acari) from the Southeast Pacific. Pac. Insects 9(4):699.

Nicholls, C. F. 1970. Some entomological equipment. 2nd ed. Can. Dep. Agric. Inf. Bull. No. 2. 85 pp.

Peterson, A. 1949. A manual of entomological equipment and methods. Ohio State University, Columbus, Ohio.

Peterson, A. 1959. A manual of entomological techniques. How to work with insects. 9th ed. Edward Brothers Inc., Ann Arbor, Mich. 435 pp.

Raw, F. 1955. A flotation extraction process for soil microarthropods. Pages 341-346 *in* D. K. McE. Kevan, ed. Soil zoology. Butterworth & Co. Ltd., London.

Richards, W. R. 1964. A short method for making balsam mounts of aphids and scale insects. Can. Entomol. 96:963-966.

Sabrosky, C. W. 1966. Mounting insects from alcohol. Bull. Entomol. Soc. Am. 12(3):349.

Salmon, J. T. 1946. A portable apparatus for the extraction from leaf mould of Collembola and other minute organisms. Dom. Mus. Rec. Entomol. (Willington) 1:13-18.

Salt, G., and Hollick, F. S. J. 1944. Studies of wireworm populations. 1. A census of wireworms in pasture. Ann. Appl. Biol. 31:53-64.

Sheals, J. G., and Hyatt, K. H. 1964. Some problems of collecting in remote areas. 1st Int. Congr. of Acarology, Ft. Collins, Colo., 2-7 Sept., 1963. Proc. Extrait de Acarologia Published in Oct. 1964. VI (fascicule hors série) :198-207.

Singer, G. 1964. A simple aspirator for collecting small arthropods directly into alcohol. Ann. Entomol. Soc. Am. 57(6):796-798.

Singer, G. 1967. A comparison between different mounting techniques commonly employed in acarology. Acarologia IX(3):475-484.

Southwood, T. R. F. 1966. Ecological methods with particular reference to the study of insect populations. Methuen & Co. Ltd., London. 391 pp.

Strandtmann, R. W., and Wharton, G. W. 1958. Manual of mesostigmatid mites parasitic on vertebrates. Contrib. Inst. Acarology. Univ. Maryland 4:330.

Teskey, H. J. 1962. A method and apparatus for collecting larvae of Tabanid (Diptera) and other invertebrate inhabitants of wetlands. Proc. Entomol. Soc. Ont. 92(1961):204-206.

Townes, H. 1962. Design for a Malaise trap. Proc. Entomol. Soc. Wash. 64(4):253-262.

Townes, H. 1972. A light-weight Malaise trap. Entomol. News 83:239-247.

Vockeroth, J. R. 1966. A method of mounting insects from alcohol. Can. Entomol. 98:69-70.

Watson, G. E., and Amerson, A. B. 1967. Instructions for collecting bird parasites. Smithson. Inst. Mus. Nat. Hist. Inf. Leaflet 477:11.

Williamson, K. 1954. The Fair Isle apparatus for collecting bird ectoparasites. Br. Birds 47:234-235.

Wilson, B. H., Tugwell, N. P., and Burns, E. C. 1966. Attraction of Tabanids to traps baited with dry ice under field conditions in Louisiana. J. Med. Entomol. 3(2):148-149.

Woodring, J. P., and Blum, M. S. 1963. Freeze-drying of spiders and immature insects. Ann. Entomol. Soc. Am. 56:138-141.

Index

Acalyptratae 152, 153
Acari (mites) 34, 38, 42, 45, 47, 49, 61, 62, 67, 68, 69, 70, 71, 73, 82, 100, 107–108, 162–166
  larva 164
Adelgidae 135
adhesives
  Canada balsam 82, 129, 133, 135, 136
  cellulose 95
  plastic 82, 95, 150
  shellac 82, 95
  shellac gel 82, 94, 150, 160, 171
  water-soluble 82, 145
Agelenidae 168
alderflies *see* Megaloptera
Aleyrodoidea 135–136
Anisoptera (dragonflies) 132
Anoplura (sucking lice) 67–69, 129–130
antlions *see* Neuroptera
ants *see* Formicidae
Aphididae (aphids) 21, 135, 159
Aphidoidea 135, 136
aphids *see* Aphididae
Apoidea 151
Araneae (spiders) 31, 34, 38, 45, 47, 71, 73, 167
Araneidae 168
Asilidae 153
aspirator
  bulb or Singer 45, 162
  general purpose 15, 28, 33, 35, 42, 43, 65, 134, 147, 148, 152, 154, 167, 168
assembling 140, 153
Astigmata 107, 166
bait
  artificial or chemical 22, 32, 154
  natural 20, 31–32, 138, 140, 153
barking 42
beating
  funnel 38
  method and application 34–38, 42, 130, 135, 167
  sheet 34, 35, 42
  tray 35
beetles *see* Coleoptera
Berlese funnel 32, 47, 51, 54, 56, 57, 68, 70, 124, 125, 128, 133, 143, 151, 161, 163, 164, 167, 168
"berlesing" 148, 153
bethyloids *see* Microhymenoptera
bird nests 47, 153, 161
biting lice *see* Mallophaga
blow flies 153, 154, 159
Bombyliidae 152
Brachycera 156

braconids *see* Ichneumonoidea
bugs *see* Heteroptera
butterflies *see* Lepidoptera
caddisflies *see* Trichoptera
calliphorids *see* blow flies
camel (cave) crickets   125
Canadian National Collection   114
Cecidomyiidae   157
Ceratogonidae   155
chalcidoids *see* Microhymenoptera
Chamaemyiidae   159
chiggers   68, 69
Chironomidae (midges)   152, 153, 154, 155, 157, 158
Cicadellidae   134, 135
Cicadidae   134
cleaning and restoring specimens   81, 82, 144, 145, 155
cleaning fluids
  ammonia   81, 82, 144, 145
  benzene   145
  carbon tetrachloride   145
  detergent   82, 160
  ethyl acetate   82
clearing agents
  Andre's fluid   169
  cedarwood oil   106, 157
  chloral hydrate   165, 169–171
  clove oil   106, 133
  KOH (caustic potash)   102, 104, 125, 128, 129, 133, 137, 157, 161, 170
  lactic acid   107, 135, 165
  lactophenol   107, 108, 125, 129, 136, 161, 166
  Nesbitt's fluid   107, 108, 166, 171
  oil of wintergreen   106
  xylol   106, 142
Clubionidae   168
Coccoidea (scales)   135–136
cockroaches *see* Dictyoptera
Coleophoridae   93
Coleoptera (beetles)   16, 21, 23, 28, 34, 42, 46, 47, 54, 58, 67, 74, 78, 79, 83, 105, 142–145
collecting bottle   42, 45
Collembola   31, 47, 125
Corrodentia (psocids)   128
Cryptostigmata (Oribatei)   107, 162, 165, 166
Culicidae (mosquitoes)   30, 158
cynipoids *see* Microhymenoptera
damselflies *see* Zygoptera
data labels   84, 99, 109–112, 126, 148, 161, 164
debris, extracting arthropods from   47–56
Dermaptera   125–127
Dictynidae   168
Dictyoptera (mantids, cockroaches)   127
Diplura   47, 125
dipper   61
Diptera (flies)   23, 30, 46, 47, 67, 152–160
dobsonflies *see* Megaloptera

double mounts   90, 156
dragonflies (Anisoptera) *see* Odonata
dredges
   Birge-Ekman   61, 65
   drag   66
   grap-and-drag   63
   Peterson   65
   Riedl   63
dredging   61–63, 65–66, 130, 143
Dri-die   67
drying agent, silica gel   84
dry preservation
   by embedding   101
   by inflating larvae   101, 147
   by freeze-drying   102, 156
ectoparasites   62, 67, 68, 128, 129, 159, 161, 164, 167
emergence cages   40, 47, 61, 66
empidids   156
Ephemeroptera (mayflies)   62, 130–131
   nymphs   130
Erigonidae   59
evisceration of large specimens   126
exuviae (cast-off larval skins)   63, 73, 132, 142, 159–160
Fair Isle apparatus   68
fixatives
   Brasil's fluid   169
   Carnoy's   159, 170
   Carnoy's fluid   170
   dioxane   104
   formal-acetic mixture   170
"flagging"   68
fleas *see* Siphonaptera
flies *see* Diptera
flotation method (treading)   43, 54, 56, 163
Formicidae (ants)   151
formulas   169–171
freezing specimens   84, 156–157
galls   42, 73, 146, 147, 158
gallflies *see* Hymenoptera
gelatin capsule   73, 148, 149, 160
genitalia, mounting and preparing   84, 129, 137, 142, 157
Geometridae   141, 142
geometrids *see* Geometridae
Gnaphosidae   168
Gracillariidae   93
grasshoppers *see* Orthoptera
Hemiptera   47, 83, 133
hepialids   140
Heteroptera (true bugs)   23, 42, 62, 67, 135–136
Homoptera (aphids, leafhoppers, scales)   23, 134–136
Hydrachnellae   165
Hymenoptera (ants, bees, sawflies, wasps, gallflies, and allies)   23, 46, 47, 74, 82, 146–151
   rearing   146
ichneumonflies *see* Ichneumonoidea

Ichneumonoidea (braconids, ichneumonflies)   146, 148
Isopoda   47
Isoptera (termites)   128
Ixodides (ticks)   67, 68, 167
   larva   167
   nymph   167
killing agents
   alcohol   16, 20, 27, 124, 134, 165, 167
   ammonia   74
   benzene   74
   carbon tetrachloride   20, 74
   chloroform   28, 62, 68, 74
   cyanide   27, 28, 74, 126, 132, 134, 136, 137, 141, 146, 153
   dichlorvos   28, 154
   ether   62, 74
   ethyl acetate   28, 74, 76, 80, 134, 143
   ethylene dichloride   74, 76, 153
   ethylene glycol   21, 28, 168
   Formalin   30, 57
   general   15, 16, 45, 73–76
   KAAD   159, 170
   tetrachloroethane   18, 74–76
killing bottle   15, 28, 33, 40, 45, 69, 73, 76, 77, 143, 147
lacewings *see* Neroptera
layering   85
leafmining insects   42, 46, 56, 57, 158
leafrolling insects   42
Lepidoptera (butterflies, moths)   14, 16, 23, 32, 46, 74, 78, 82, 91, 93, 138–142
   rearing   138, 140
light
   collecting at   15–20, 130, 137, 138, 139, 143, 152, 153
   mercury-vapor   15, 18
Linyphiidae   168
Lycosidae   168
Lyonetiidae   93
Mallophaga (biting lice)   67, 128
mammal nests   43–47, 143, 153, 161
mantids *see* Dictyoptera
mayflies *see* Ephemeroptera
Mecoptera (scorpionflies)   137
Megaloptera (dobsonflies, alderflies)   136
Membracidae   134
Mesostigmata   107, 164, 165
Microhymenoptera (cynipoids, proctotrupoids, chalcidoids, bethyloids)   106, 146, 148, 149
Microlepidoptera   138, 141
microscopes
   binocular stereo-   122
   compound   123
   drawing attachments for   123
   lamp for   123
   scanning electron   123
microscopical preparation
   clearing   104
   dehydration and hardening   106, 107, 129, 136

   dipterous   157
   removal of soft parts   102
   staining   103, 129
   temporary mounts   107, 125
midges *see* Chironomidae
Miridae   134
mite-brushing machine   38, 164
mites *see* Acari
mites associated with invertebrates   69, 70
mold preventatives
   Beechwood creosote   115
   ethyl acetate   80
   naphthalene   80
   paradichlorobenzene   80
   phenol   80
mosquitoes *see* Culicidae
moths *see* Lepidoptera
mounting media
   Berlese's   105, 125
   Canada balsam   104, 105, 129, 130, 133, 135, 136, 161
   C-M (methyl-cellulose) fluid   109
   de Faure's   105, 125, 170
   glycerin jelly   105, 109
   glycerol   109
   gum-chloral   109
   Hoyer's   70, 105, 108, 125, 133, 135, 166, 170
   polyvinyl lactophenol   105, 171
mounting beetles on cards   144
muscids   154
Myriapoda   47
Nematocera   152, 155, 157
Nepticulidae   93
nets
   aquatic   58, 143
   Birge cone   61
   dip   61
   dredging   143
   flying insect   14–15, 132, 137–139, 146, 152
   general purpose   11–12, 125, 137–139, 152
   plankton   58, 65
   single cone   65
   sweep   34, 152
Neuroptera (lacewings, antlions)   137
Noctuidae (phalaenids, Phalaenidae)   18, 139–141
Notonectidae   134
Odonata (damselflies, dragonflies)   62, 82, 131–132
Opiliones   47
Orthoptera (Saltatoria) (grasshoppers)   76, 125–127
Parasitengona   165
Phalaenidae *see* Noctuidae
Phasmatodea (stick insects)   127
photoeclector   54
Phylloxeridae   135
pinning
   Diptera from alcohol   156–157

forceps 88, 113
methods 85–88, 126–127, 131, 132, 134–135, 137–138, 141, 144, 145, 146–147, 148, 150–151, 155, 160
pins
  choice of 87–88, 134, 141, 144, 148, 155
  minuten nadeln 88, 90, 126, 141, 156
Plecoptera (stoneflies) 62, 128
  nymphs 128
pointing 85, 94–97, 134, 144, 147, 150
points 95–97, 156, 160
preservatives
  acetoalcohol mixture 169
  AGA 99, 133, 169
  alcohol (ethyl and methyl) 27, 52, 63, 99, 124, 125, 128, 129, 131, 132, 136, 137, 138, 142, 145, 147, 149, 151, 155, 159, 165, 167, 169
  ethylene glycol 21, 28, 30
  Formalin 30, 63, 99, 147
  glycerin 52, 70, 99, 103, 104, 149, 165
  Kahle's fluid 170
  Koenicke's fluid 165, 171
  lactophenol 125, 133, 136
  Oudemans' fluid 99, 165, 171
  Pempel's fluid 99, 171
  xylene 126
  proctotrupoids *see* Microhymenoptera
Prostigmata 107, 166
Protura 47, 125
Pseudoscorpionida 47
psocids *see* Corrodentia
Psocoptera 47
Psyllidae 134
rearing 70–73, 138, 140, 146, 148, 157, 165, 167
  cage 71
recording insect songs 126
relaxing 79–81, 145
  box 80
  fluids
    ammonia 81, 145
    ammonium hydroxide 81
    Barber's 81, 169
Salticidae 168
sarcophagids 154
sawflies *see* Symphyta
Scolioidea 151
scorpionflies *see* Mecoptera
searching and handpicking 42–43, 47, 58, 134, 137, 143, 146, 167
separator 41, 47
shipping specimens 118–122
sifter 32, 54, 143
sifting 43, 167
silverfish *see* Thysanura
Simuliidae 153, 154
Siphonaptera (fleas) 47, 67, 161
slide ringing substances
  Canada balsam 109

Duco enamel   105
 euparal   109
 Glyceel   109
 Glyptal electrical finish   105, 109
 gold size   105
 Murrayite   105
 Zut   109
solvents
 acetone   82
 alcohol (ethyl and methyl)   82, 95
 ethyl acetate   82
 polyvinyl alcohol   94
 xylene   82
Sphecoidea   151
Sphingidae   140
spiders *see* Araneae
spreading   81, 90–94, 136, 137, 141, 150–151
spreading board   90, 92
stage for handling pinned insects   124
stains
 acid Fuchsine   129
 Barr solution   169
 carbolfuchsin   169
 chlorazol black E   137
 lactophenol – Methyl Cotton Blue   71
Staphylinidae   144
stemminers   158
stick insects *see* Phasmatodea
stoneflies *see* Plecoptera
storage of specimens
 in cabinets   113
 in glass-topped cases   113
 in liquid   116
 in paper envelopes and tubes   82, 119, 132
 in Schmitt boxes   113
 on pins   113
 on slides   116
 protection of   115
strainer   31, 58, 61
Strepsiptera (stylops)   145
stylops *see* Strepsiptera
subimagoes, device for collecting   131
sucking lice *see* Anoplura
sugaring   32, 139, 153
sunlight   35, 132, 139, 153
sweeping   33–34, 128, 135, 137, 140, 143, 147, 148
Symphyta (sawflies)   146
 larvae   146, 147
Syrphidae   152, 154
Tabanidae   27, 152, 154, 158
Tachinidae   154, 159
tanglefoot   32
termites *see* Isoptera
Tetragnathidae   168
Theridiidae   168

Thomisidae 168
thrips *see* Thysanoptera
Thysanoptera (thrips) 47, 132–133
Thysanura (silverfish) 47, 124
"tick drag" 68
ticks *see* Ixodides
Tipulidae 152, 155
Tischeriidae 93
traps
  aquatic 63
  bait 20–22, 125, 127, 140
  benthonic 65
  box 65
  cone stream 66
  emergence 63, 64
  funnel 16, 18, 63, 65
  Herting tent 154
  lentic environment 63
  light 15, 16, 18, 130
  lotic 66
  Malaise 23, 27, 137, 146, 148, 151, 154, 156
  pan 30, 148, 155, 171
  pit-fall 21, 167
  simple cone or net 63
  submerged aquatic 63
  surface cone 63
  tent-window 28
  tow 65
  triangular stream 66
  visual-attraction 30
  window 28
treading *see* flotation method
Trichoptera (caddisflies) 62, 137
  rearing 138
ultrasonic cleaner 82
vacuum collector 38, 164, 167, 168
vertebrate hosts, collecting 67–68
Vespoidea 151
vials
  homeopathic 99
  micro- 73, 103, 137, 160
  shell 124
wasps *see* Hymenoptera
wetting agents
  detergent 30, 168
  ethylene glycol 168
Zygoptera (damselflies) 132